ISBN-13: 978-1512044706

ISBN-10:   1512044709

Self-published on Create Space

Forward

This book is chapter 1 of a series of book. The entire series is an introduction to combinatory theory, number theory, topology, groups, rings, fields, modules, algebraic and integral extensions, noncommutative algebra, algebraic number theory, algebraic geometry, algebraic topology, and even more. The title of the series is Abstract Algebra and Discrete Mathematics.

The focus is on breadth, rather than depth. Excellent books already exist for any one of these topics in detail, and I don't want to reinvent that wheel. Instead, this book knits them all together, providing a foundation for each topic in turn. By analogy, you might point your backyard telescope to every corner of the galaxy, in an effort to comprehend its scope, beauty, and diversity. You might not understand the Crab Nebula in all its detail. This is not the Hubble Space Telescope, but you should walk away with an appreciation for the vastness and the wonder of the galaxy, in this case, the galaxy of modern mathematics. If, from time to time, you find yourself saying, "How did anybody ever think of that?", then I have succeeded.

Each chapter builds on the information that has gone before, and forward references are rare, though they do happen from time to time. I hope this series of books is more accessible than a sea of disconnected web pages, which is the hallmark of most math websites.

Though not strictly necessary, it is helpful if you are familiar with introductory calculus on real and complex variables. Yes, this book addresses discrete mathematics, but it is surprising how discrete and continuous math can support each other. In fact, an entire branch of mathematics, analytic number theory, answers questions in number theory using analytic functions on the complex plane. Closer to home, chapter 1 will describe formal derivatives, the differentiation of polynomials for algebraic purposes. These polynomials might be restricted to the integers mod 7, and yet the derivative of p(x) still makes sense. Thus, a background in calculus is helpful from time to time, but not vital. If you have never seen a derivative or an integral before, you will still be able to follow 95% of this book.

This book is Chapter 1, Data Structures,

By Karl Dahlke and Kermit Rose, Copyright © 2015

Table of Contents

Properties of Arithmetic

Teachers have told you since elementary school that addition is commutative, that x+y is the same as y+x. Is this true? How do you know? If a proof is a line of reasoning that convinces you, then perhaps this will do the trick. Take a stack of plates in one hand, and a stack of plates in the other, and place one stack on top of the other. It won't matter whether you place the first stack on top of the second or the second on top of the first; you will have just as many plates.

It's sort of the conservation of stuff. This "proof" was sufficient for thousands of years, and was not replaced with a more formal proof until the twentieth century. Alfred Whitehead and Bertrand Russell co-wrote the Principia Mathematica, in 3 volumes, which builds modern mathematics from the ground up.

By definition, Zero is the empty set, 1 is the set containing 0, two is the set containing 0 and 1, 3 is the set containing 0 1 and 2, and so on. With the positive integers established, and addition defined, one can prove commutativity by mathematical induction.

The logic path would be

$0 + 0 = 0 + 0$

$0 + 1 = 1 + 0$ because both are equal to 1.

$1 + 1 = 1 + 1 = 2$

$1 + 2 = 1 + (1 + 1) = (1 + 1) + 1 = 2 + 1$

Now suppose $j < k$ and for those particular values of j and k, we have proven that

$j + k = k + j$

then we note that $(j + 1) + k = j + (1 + k) = j + (k + 1) = (j + k) + 1 = (k + j) + 1$

$= k + (j + 1)$.

To be perfectly rigorous, you have to develop induction from first principles. The reader must plow through hundreds of pages of set theory before confirming something that a 6-year-old understands, that addition is commutative. I am going to skate past the nuances of ordinal and cardinal arithmetic and simply state that your childhood intuition is correct. Addition is commutative because you can stack the plates in either order and the number of plates is the same.

Addition is also associative, that is, $(x + y) + z = x + (y + z)$. Start with three stacks of plates that I will call A B and C. You can place B on top of A, and then C on top of the A + B stack, or you can place C on top of B and put the combined pile on top of A. The number of plates is the same regardless of which of these two ways you stack them.

One can explore the properties of multiplication by baking some cookies. Place perfectly formed balls of cookie dough on a metal cookie sheet, in a grid that has 4 rows and 7 columns.

In 20 minutes, you will have 4 times 7 = 28 cookies. Rotate the sheet 90 degrees and the number of cookies does not change. Thus 4×7 = 7×4.

The distributive property is also demonstrated by cookies in a grid. Draw a vertical line between the second and third columns. This generates the equation 4×7 = 4×2 + 4×5. In general, $x \times (y + z) = x \times y + x \times z$. This is the distributive property, and it works just as well from the other side: $(x + y) \times z = x \times z + y \times z$.

Finally, multiplication is associative. Arrange sugar cubes, or building blocks, in a box that is 3 by 4 by 7. The floor of the box contains 3×4 or 12 blocks, and there are 7 layers, hence 84 blocks. Lay the box down on its front face and the floor of the box contains 4×7 or 28 blocks. There are 3 layers, hence 84 blocks. Therefore $(x\ y)\ z = x\ (y\ z)$.

These properties extend, in a natural way, to real numbers. When placing stacks of plates on top of one another, pretend like the top plate is broken. Thus 3.5 + 4.7 is the same as 4.7 + 3.5. In the same way, let some of the cookies, or blocks, be incomplete. The properties of integers extend to fractions, and then to real numbers by continuity. This is all intuitive, but the good news is, all of it is true, and can be verified from first principles.

Modular Math

If a movie starts at 11:00 and lasts 2 hours, when does it end? It ends at 1:00 of course. This is a world, a circular world if you will, or a ring, where 11+2 = 1. Mathematicians use these rings all the time, but by convention, they begin with 0. Thus, in Clock arithmetic, 7+5 = 12, and 7+6 = 1.

Arithmetic mod 12 is a generalization of clock arithmetic. However, in mod 12, we restrict ourselves to the numbers 0, 1, 2, 3, 4, 5, 6, 7, 8, 9, 10, and 11. Thus in mod 12 arithmetic, 0 is the additive identity, while in clock arithmetic, 12 is the additive identity.

Subtraction is also defined mod 12, just march around the circle in the opposite direction. 3-7 = 8 mod 12.

Multiplication is repeated addition, 5×7 is 7+7+7+7+7, so this too is well defined. Even division is defined in some cases. 3/7 = x, where x is some number from 0 to 11 that when multiplied by 7 yields 3. Fish around and come up with x = 9. But other quotients are not well defined, such as 5/3. Anything times 3 is going to become 0 3 6 or 9, not 5. We'll explore modular division later on, once we have a little more machinery in place.

Homomorphism

The homomorphism is perhaps the most important idea in abstract algebra. You'll run into it repeatedly. The word is Greek for having the same shape. A homomorphism is a function from one structure to another that preserves certain aspects of the structure.

Modular mathematics is a perfect example. The set upstairs is the integers, and the set downstairs is the numbers mod 5, consisting of 0 1 2 3 and 4. The function, or map, divides n by 5 and takes the remainder. This becomes the image of n downstairs, also known as f(n). For example, 22, 357, 262, 982, and -8 all map to 2. The reason for a 0 based system now becomes clear. The remainder, after dividing by 5, is a number from 0 to 4, rather than 1 through 5, so best to start with 0. Thus under the mod 5 homomorphism, 10 and 595 both map to 0.

All the numbers that map to 0, i.e. the multiples of 5, form the kernel of the homomorphism. This is the heart of the homomorphism.

To be a homomorphism, the function must preserve addition and multiplication. Let f and g be homomorphic functions. Let x and y be integers upstairs, and let u and v be the corresponding numbers downstairs. In other words, u = f(x) and v = f(y). A homomorphism requires f (x + y) = u + v = f(x) + f(y), and f (x y) = u v = f(x) f(y). You can add or multiply upstairs, then drop down, or you can add or multiply downstairs; the result is the same. The simpler structure downstairs captures at least some of the essence of the more complicated structure upstairs, and in some cases the structure downstairs is easier to analyze. That is the power of the homomorphism.

How do we know that modular map is a homomorphism? Start with (5 x + 2) and (5 y + 4) upstairs, where x and y are any integers. Naturally (5 x + 2) maps to 2 and (5 y + 4) maps to 4. Now take the sum, upstairs and down. Upstairs we have ((5 x + 5 y) + (2 + 4)), which is 5 (x + y + 1) + 1, which is 1 + some multiple of 5. The image of the sum is 1. Downstairs, we observe that 2+4 = 1. It's the same. The function respects addition. How do you know that 5 x + 5 y is a multiple of 5? Because of the distributive property: 5 × x + 5 × y = 5 × (x + y). Or you can just say that clumps of 5 plus clumps of 5 produces more clumps of 5. These are some of the low level details that I said I would skip, and I will, but it's good to have an intuition for what is going on.

Multiplication runs the same way. (5 x + 2) × (5 y + 4) = 25 x y + 20 x + 10 y + 8, which is 3 plus some multiple of 5. Downstairs, we observe that 2 × 4 = 3. The map respects multiplication. It is indeed a homomorphism.

Instead of talking about multiples of 5, use the letter m instead. The reasoning is the same "mod m". The function "mod m" is the remainder after dividing by the integer m.

Here m could be 5 or 12 or 173 or 6 million or anything you like. The function mod m, which is dividing by m and taking the remainder, is a valid homomorphism that respects addition and multiplication. The kernel consists of the multiples of m.

Set m = 2 and find the parity function, where n is declared even or odd. Even numbers map to 0 and odd numbers map to 1. This reproduces some tables that you may have seen before.

even + even = even

even + odd = odd

odd + odd = even

even times even = even

even times odd = even

odd times odd = odd

Instead of the integers, place the real numbers upstairs and once again let m = 5. The remainder of r mod 5 is now a real number between 0 and 5, including 0 and excluding 5. The map is still a homomorphism, respecting addition and multiplication, and the proof is exactly the same. Other values of m are possible, such as 2.5 or pi. If m = pi, then "mod m" determines the circumference of the circle downstairs. Wrap the real line around the circle, like thread around a spool, repeatedly.

The following example reduces the real numbers mod 1, e.g. 53.27 becomes 0.27. In the illustration, think of x as 0.27, although x could be any real number in the interval [0,1). The use of the left bracket, [, shows that 0 is included. The use of the right parenthesis, ")", shows that 1 is not included. This is called a half-open interval, and it represents the range of the "mod 1" homomorphism.

$$\ldots \text{-}4 + x \quad \text{-}3 + x \quad \text{-}2 + x \quad \text{-}1 + x \quad 0 + x \quad 1 + x \quad 2 + x \quad 3 + x \quad 4 + x \ldots$$

Technically, the operations of a homomorphism, upstairs and down, need not be the same. Thus, it is perfectly reasonable to make the upstairs operation look like addition and the downstairs operation look like multiplication. Consider the exponential map r --> 4^r. Remember that $4^{(a+b)} = 4^a$ times 4^b. Thus addition upstairs corresponds to multiplication downstairs. We can't concisely say, "f respects addition", but we can say "f is a homomorphism from the reals under addition to the positive reals under multiplication."

-2 --> .0625
-1.5 --> .125
-1 --> .25
-.5 --> .5
0 --> 1
.5 --> 2
1 --> 4
1.5 --> 8
2 --> 16
2.5 --> 32
3 --> 64
3.5 --> 128

In this case the map can be reversed, going from the positive reals back to the reals, giving a function that is known as log base 4. This is a reverse homomorphism, changing multiplication back to addition.

.0625 --> log4(.0625) = -2

.125 --> log4(.125) = -1.5

.25 --> log4(.25) = -1

.5 --> log4(.5) = -1/2

1   --> log4(1) = 0

2   --> log4(2) = .5

4   --> log4(4) = 1

8   --> log4(8) = 1.5

16   --> log4(16) = 2

Casting out Nines

Without a calculator, quickly determine whether 12485240113 is divisible by 9.

The trick is to add the digits, casting out 9's along the way.

$1 + 2 = 3$

$3 + 4 = 7$

$7 + 8 = 15$

$15 - 9 = 6$

$6 + 5 = 11$

$11 - 9 = 2$

$2 + 2 = 4$

$4 + 4 = 8$

$8 + 0 = 8$

$8 + 1 = 9$

$9 - 9 = 0$

$0 + 1 = 1$

$1 + 3 = 4$

12485240113 is not divisible by 9. The remainder is 4.

Why does this trick give the correct answer?

Write this 11-digit number as a sum of individual digits times powers of ten.

12485240113

$= 1 \times 10^{10} + 2 \times 10^9 + 4 \times 10^8 + 8 \times 10^7 + 5 \times 10^6 + 2 \times 10^5 + 4 \times 10^4 + 0 \times 10^3 + 1 \times 10^2 + 1 \times 10 + 1$

When working mod nine, ten is the same as one. Ten to any power is also the same as one.

$10 = 9 + 1$

$10^2 = 11 \times 9 + 1$

$10^3 = 111 \times 9 + 1$

Etc

So we can simply add up the digits and see if that is divisible by nine. The result is 31, and $3+1 = 4$, hence 4 mod 9. Our original number is not divisible by 9.

This procedure is sometimes called "casting out nines", because nines can be discarded as you go. They're always divisible by 9 anyways. Is 119979 divisible by 9? Cast out the nines and add up the remaining digits, giving $1 + 1 + 7 = 9$. Cast out this final 9 and yes our original number is divisible by 9.

This works because we are using a base ten number system. If you usually write your numbers in octal, i.e. base 8, you can tell very quickly whether a number is divisible by 7. Add up the octal digits and see if the result is divisible by 7.

Since 3 goes into 9, you can add up the digits to determine whether a large number is divisible by 3.

Since 2 and 5 divide into 10, you only need consult the last digit to see if a number is divisible by 2 or 5, or 10. Examine the last two digits to see if a number is divisible by 4, 20, 25, 50, or 100.

Add and subtract the alternating digits of a number to see if it is divisible by 11. This works because 1 is 1 mod 11, 10 is -1 mod 11, 100 is 1 mod 11, 1000 is -1 mod 11, and so on.

Mod 9 is sometimes used as a simple form of error detection. Multiply 37 by 29 with pencil and paper, and get 1073. Check your work by casting out nines. It's a homomorphism - so the answer mod 9 is the same as doing the whole problem mod 9, which is much easier. 37 becomes 1, and 29 becomes 2, thus the product should be 2 mod 9, and indeed it is. $1+0+7+3 = 11$, and then $1+1 = 2$.

Monomorphism, Epimorphism, Isomorphism

Mono is Greek for "one", hence a monomorphism is a homomorphism where at most one element upstairs maps to each element downstairs. There is no overlap. It is also called an embedding, or an injective function. Mod m is not a monomorphism, because many different numbers upstairs map to the same thing downstairs. 7 and 12 both become 2 mod 5. However, the exponential function 10^r is a monomorphism. Take the log to find the unique r upstairs that leads to any positive value below. $10^{\log(r)} = r$ for $r > 0$.

A homomorphism f is a monomorphism iff the kernel contains exactly one element. (Note that I use "iff" to mean "if and only if"; it is not a misspelling.) When you see A iff B, you will need to prove A implies B, and then B implies A. Thus, f is a monomorphism exactly when f has a trivial kernel. By trivial kernel, I mean that only the identity element upstairs maps into the identity element downstairs. For addition, the identity element is 0, and for multiplication, the identity element is 1.

For notational convenience, let I represent the identity element of any group. Thus if the group is an additive group, I = 0. If the group is a multiplicative group, I = 1. Then f(I) = I.

Example multiplication mod 5 is equivalent to addition mod 4.

The elements of addition mod 4 are I, 1, 2, and 3.

Let the elements of multiplication mod 5 be I, 2, 3, 4.

Choose one of the mappings {I, 1, 2, 3} --> {I, 2, 4, 3} or {I, 1,2, 3} --> {I, 3, 4, 2}.

We illustrate using the first mapping. f(I) = I, f(1) = 2, f(2) = 4, f(3) = 3.

2+3 = 1, in addition mod 4

--> f (2) x f (3) in multiplication mod 5

--> 4 x 3 = 2, in multiplication mod 5

2 + 3 = 1, in addition mod 4

--> f (2) x f (3) in multiplication mod 5

--> 4 x 2 = 3 in multiplication mod 5

We use the ? symbol to indicate addition or multiplication, which ever applies. Also, define @x to be the element such that @x ? x = I. @x is called the inverse of x.

The kernel is the preimage (reverse map) of I, and it contains at least I. If f is a monomorphism then the preimage is unique, and the kernel contains only I. Conversely, suppose x and y upstairs both map to v downstairs. Then @x?y maps to @v?v, which is I. Two things upstairs map to the identity downstairs and the kernel is more than just I. Thus the kernel becomes a simple test for an embedding.

This is valid for groups, rings, and modules, which is 99% of what we care about, but there is a catch. In a monoid, which is only half a group, you might not be able to subtract. There might be no x-y or x/y. Consider the following monoid homomorphism. Let f map the positive integers to the number of prime factors, including repeated primes, in each integer. Thus f(1) = 0, f(7) = 1, f(49) = 2, f(30) = 3, f(17) = 1, and so on. Multiplication upstairs corresponds to addition downstairs; we have a homomorphism. The kernel is unique; only 1 maps to 0. But every prime maps to 1, hence f is not a monomorphism. The proof breaks down because x/y is not part of the domain upstairs. But this is rather atypical; for groups rings and modules, f is a monomorphism iff its kernel is a single element.

An epimorphism is a homomorphism that covers the entire structure downstairs. Nothing is left out in the cold. If 10^r maps real numbers into real numbers, it is a monomorphism (as described above), but not an epimorphism, because the negative numbers are not touched. However, restrict the range downstairs to the positive reals, and the map becomes an epimorphism again.

The prefix iso is Greek for the same; thus an isomorphism is a homomorphism between structures that are essentially the same. It is both a monomorphism and an epimorphism. Everything downstairs has one and exactly one preimage upstairs. The map can be reversed. The two structures are indistinguishable; one is merely a relabeling of the other.

Before calculators became small and affordable, engineers carried slide rules around in their pockets. Why? Because the slide rule implements a convenient isomorphism, namely log base 10. One could multiply two numbers by adding their logs, then raising 10 to that power. Addition was easily accomplished with pencil and paper, and soon the product appeared without much fuss. Divide the log by 2 to find the square root, and so on. In the movie Apollo 13 you can see a room full of engineers with their trusty slide rules, performing some rapid fire calculations to see if the injured space craft is still on course for Earth. This illustrates the value of an isomorphism. Sometimes a problem can be transformed into another domain where it is easier to solve; then the solution can be brought back home.

An automorphism (for self-shape) is an isomorphism from a structure onto itself. Replace x with cx, for any nonzero constant c, and find an automorphism on the reals that respects addition. $c(x+y) = cx + cy$, and every x has a unique preimage, namely x/c. This doesn't respect multiplication, but it is an automorphism for addition.

A more complete automorphism, respecting both addition and multiplication, is conjugation in the complex plane, mapping i to -i. This is reflection through the x axis. Visualize this mapping as follows. Draw a horizontal line to represent the real numbers. Through the zero point, draw a vertical line to represent the imaginary numbers. This is very much like the xy plane, but x is the real component of a complex number and y is the imaginary component. Pick a point in the first quadrant. This is a point above the real axis and to the right of the imaginary axis. Suppose you pick the point(4,2), which represents 4+2i. The automorphism takes the point 4+2i to the point 4-2i. The complex number 4+2i is the point 4 units to the right of the imaginary axis and 2 units above the real axis. Under this automorphism, it will be mapped to the point 4 units to the right of the imaginary axis and 2 units below the real axis. This automorphism, complex conjugation, is reflection through the real axis, and it respects both addition and multiplication.

For all real numbers, a, b, define conjugate (a + b i) to be equal to (a – b i). Then conjugate ((a +b i) + (c+ d i)) = conjugate ((a + c) + (b + d) i)) = ((a + c) – (b + d) i) = (a – b i) + (c - d i) = conjugate(a + b i) + conjugate(c + d i)

Conjugate ((a + b i) x (c + d i)) = conjugate ((a c – b d) + (a d + b c) i) = (a c –b d) - (a d + b c) i = (a – b i) x (c – d i) = conjugate (a + b i) x conjugate (c + d i)

An endomorphism, (the shape inside), is a homomorphism that maps a structure into itself. An example is x --> 2x, carrying the integers into themselves. This map is a homomorphism under addition, but does not cover all of the integers, so cannot be considered an automorphism.

The composition of monomorphisms is a monomorphism.

The composition of epimorphisms is an epimorphism.

The composition of isomorphisms is an isomorphism. This follows from the previous two.

The composition of endomorphisms is an endomorphism.

The composition of automorphisms is an automorphism.

Discrete Logs

Let $f(n) = 2^n$ mod 11. Thus, $f(0) = 1, f(1) = 2, f(2) = 4, f(3) = 8, f(4) = 5, f(5) = 10$, and so on. This is an exponential function. Hence, it implements a homomorphism from the integers mod 11 onto the nonzero integers mod 11. Addition upstairs corresponds to multiplication downstairs. Since 2^{10} takes you back to 1, we can restrict the domain upstairs to the integers mod 10 and find an isomorphism. Addition mod 10 is the same as multiplication of the nonzero elements mod 11.

Here is the multiplication table mod 11, shown in context of addition mod 10.

Note that the pattern in this multiplication table is the same pattern seen in the addition mod 10 table. For example: look at 9+9 mod 10 = 8 mod 10.
$f(9) = 6$, and $f(8) = 3$. Note that 6 x 6 mod 11 = 36 mod 11 = 3 mod 11.

		0	1	2	3	4	5	6	7	8	9
		1	2	4	8	5	10	9	7	3	6
n	f(n)										
0	1	1	2	4	8	5	10	9	7	3	6
1	2	2	4	8	5	10	9	7	3	6	1
2	4	4	8	5	10	9	7	3	6	1	2
3	8	8	5	10	9	7	3	6	1	2	4
4	5	5	10	9	7	3	6	1	2	4	8
5	10	10	9	7	3	6	1	2	4	8	5
6	9	9	7	3	6	1	2	4	8	5	10
7	7	7	3	6	1	2	4	8	5	10	9
8	3	3	6	1	2	4	8	5	10	9	7
9	6	6	1	2	4	8	5	10	9	7	3

Like any exponential isomorphism, this can be reversed. The inverse operation is called a discrete log. It is a log in the traditional sense, but the inputs are discrete integers rather than a continuous flow of real numbers. The log of 5 is 4, because 24 = 5 mod 11. If it is not clear from context, then the base and the modulus must be explicitly stated. For instance, the discrete log of 5, mod 11, base 2, is 4, because $2^4 = 5$ mod 11.

n mod 10	0	1	2	3	4	5	6	7	8	9
2^n mod 11	1	2	4	8	5	10	9	7	3	6
log2(2^n) mod11	0	1	2	3	4	5	6	7	3	6

The exponential map is very efficient, even for large numbers. Let m be a 50 digit number, and let x be a 30 digit number, and consider x raised to the million and 3 mod m. You don't really want to multiply x a million times over, and you don't have to. Write the exponent 1,000,003 in binary, giving 11110100001001000011. This binary representation tells you that 1,000,003 is:

$$((((((((((((((((((1\times2+1) \times2+1) \times2+1) \times2+0) \times2+1) \times2+0) \times2+0) \times2+0) \times2+0) \times2+1) \times2+0) \times2+0) \times2+1) \times2+0) \times2+0) \times2+0) \times2+0) \times2+1) \times2+1$$

Since this serves as an exponent, each ×2 is a square, and each +1 becomes multiplication by x. The algorithm reads the binary number right to left. The last 1 in this number represents x to the first, and that has to be part of the product. The next 1 over is x^2, which is easily calculated. Call it s. With a 1 in the second position, this is also part of the product. Multiply s by x to get x^3, and call that result t. Next replace s with s^2, which is actually x^4. (Remember to reduce mod m at each step.) There is a 0 in the corresponding slot of the binary exponent, so keep going. Square s again, and again, and again, and again, giving x^{64}. This is part of the product, because there is a 1 in the corresponding slot, so multiply t by s. Continue all the way to x^{524288}, and t holds the answer. Here is some pseudocode for the algorithm. All variables are high precision integers.

```
/* compute b to the e mod m */

ModularExponentiation(b, e, m)

        {a = 1; /* accumulator / answer */

        s = b;

        while(e > 0) {

                if(e is odd) a = a×s mod m;

                s = s×s mod m;

                e = e/2; /* shift the exponent down */                    }

        return a; }
```

Unfortunately, there is no efficient procedure for finding a discrete log. 37 to what power = 400 mod 1031 - there really isn't any way to know except to try each value of x in turn. These "one-way" functions are useful for cryptography. It is easy to mathematically encode the message, but impossible to decode it (without some sort of secret key). We will say more on this later.

Binary Search

A forward function can always be reversed if it is monotonic, i.e. a larger input leads to a larger output. Consider the example of the real exponential function 10^x. When x gets bigger, 10^x gets bigger. f(1) = 10, f(2) = 100, f(3) = 1,000, and so on. Even in between, $10^{1.7}$ is larger than $10^{1.6}$, is larger than $10^{1.58}$. Assume there is an efficient procedure for calculating 10^x, and now we want to find x such that 10^x = v. In other words, we are looking for the log of v. There is actually a formula from calculus for the log of v, but let's set that aside for the moment. If v has 3 digits to the left of the decimal point, e.g. v = 321.67, then make two guesses for x, namely 2 and 3. Since 10^2 = 100, and 10^3 = 1000, the first is too small and the second is too large. Let L = 2, the left boundary, and let R = 3, the right boundary. Let m = (L+R)/2. Thus m is the midpoint of our segment that contains the magical value x. Compare 10^m with v. If 10^m is too small then make m the new left boundary; otherwise make m the new right boundary. Repeat this until R-L is small, whence x, trapped in between, has the desired precision. Here is some code for a binary search, turning a forward monotonic function into a backward function. All variables are floating point, with at least 6 digits of precision.

Find x such that forward(x) = v, x = backwards(v).

```
{

    /* make an intelligent guess for L and R */

    /* this will depend on the forward function and on v */

    /* for illustration, let's start with the range 0 to 100 */

    L = 0;

    R = 100;

    /* stop at 6 digits of precision */

    precision = 0.000001;

    while(R - L > precision) {

            m = (L+R) / 2;

            if(forward(m) < v) L = m; else R = m;

    }

    return L; }
```

Unlike real exponentiation, modular exponentiation is not monotonic. A larger input does not lead to a larger output. In fact the output looks almost random relative to the input. The results are scrambled, so binary search cannot home in on the correct value. In modular arithmetic, the exponential function cannot be reversed.

To illustrate, here is 11^x mod 31, as x runs from 0 to 30. Notice that the output jumps all over the place.

1 11 28 29 9 6 4 13 19 23 5 24 16 21 14 30 20 3 2 22 25 27 18 12 8 26 7 15 10 17 1

Function Composition

Consider the unit disk in the xy plane. This is the set of points on and inside the circle centered at the origin with radius one. It is the set of points satisfying $x^2 + y^2 = 1$. Let f, g, and h be 3 functions that map the unit disk into itself. It doesn't much matter what these functions are. Let's make up some functions just to illustrate the concept. A point on the disk is represented by two numbers. The distance of that point from the origin is one of the numbers. The name of that distance is "radius". The other number is the angle between the x-axis and a line connecting that point to the origin. We have described what is called polar coordinates. Our functions, f, g, and h will each be functions with two arguments.

Let f rotate the disk 60 degrees clockwise. Let g reflect the disk through the y axis like a mirror. Let h pull the disk towards the origin and stretch it away from the edge, e.g. r^2, where r is the radius. Composition is a fancy word for applying one function and then another. Thus (f compose g) is f followed by g. In other words, (f compose g) (x) = g(f(x)). A point p maps through f, and through g, and winds up somewhere else in the disk. The combined function is written f x g, or simply fg.

In general, composition is not commutative. Start with a dot at 12:00, at the top of the circle. Call this point p. The radius of p is 1 because it is on the unit circle. Its angle is 90 degrees. Apply f to move the point to 4:00, and then g to move p to 8:00. However, gf moves p first to 12:00, keeping it where it is, and then to 4:00. The two functions are not the same.

Function composition is however associative. Take any point p in the disk and run it through fg, and then through h. This is, by definition, f followed by g followed by h. Next run p through f, and then through the composite function gh. This is again f followed by g followed by h. Therefore (f x g) x h = f x (g x h). We can write fgh without any ambiguity.

Permutations

A real number can indicate a point on a line. Two numbers might designate a point in the xy plane. Three numbers could fix a point in space. But sometimes numbers define an action. Sometimes numbers tell you how things move. This is the case when numbers describe a permutation.

Imagine 6 stuffed animals sitting on a shelf. These are ape, bear, cat, dog, eagle, and fox. How many ways can they be arranged left to right? Choose any of the 6 animals, and place it in the leftmost position. Say it's the cat. Then place any of the 5 remaining animals next to the cat, perhaps the fox. Next, there are 4 animals to choose from, then 3, then 2, then 1. Therefore, the number of arrangements is $6 \times 5 \times 4 \times 3 \times 2 \times 1 = 720$. This is also written 6! and is called 6 factorial.

A permutation is the act of moving these animals about. For example, shift all the animals to the left and put the leftmost animal on the right. This is called a circular shift, and it is one of many possible permutations. Do this 6 times and you are back where you started.

A	B	C	D	E	F
B	C	D	E	F	A
C	D	E	F	A	B
D	E	F	A	B	C
E	F	A	B	C	D
F	A	B	C	D	E
A	B	C	D	E	F

A permutation is defined by a sequence of 6 numbers which represent instructions for moving the 6 animals about. There are two conventions to choose from. Consider the sequence 234561. This might mean that the animal in position 1 moves to position 2, 2 moves to 3, 3 moves to 4, and so on until 6 moves to 1. Or it could mean that the animal that was in position 2 is now moved to slot 1, position 3 moves to position 2, position 4 moves to position 3, and so on until position 1 appears in position 6. Is it a right circular shift or a left circular shift? Some books use one convention; some use the other. I will adopt the second convention for two reasons.

[1] Put a tag on each animal, 1 2 3 and so on, indicating their start positions. Then apply a permutation to the animals. If the permutation starts with 4, then the fourth animal, the dog, moves to slot 1. His tag displays 4, and now he is in the first position, just like the number 4 in our sequence. If the next number is 5 then the eagle moves into the second position and displays his tag showing the number 5. After the permutation is applied, the tags display the same numbers as the permutation itself. Put another way, when the permutation is applied to 123456

(start), the result is exactly the same as the permutation. You can read a sequence of numbers as a formula for moving the animals about, or the way they will look after they have been moved, if they were in their "start" positions. The permutation 426135 moves animal 4, the dog, to position 1. It moves animal 2, the bear, to position 2. It moves animal 6, the fox, to position 3. It moves animal 1, the ape, to position 4. It moves animal 3, the cat, to position 5. Finally, it moves animal 5, the eagle, to position 6.

[2]. Let f be a permutation and let g be another permutation. These are really functions from a set of 6 elements into itself. This point goes here, this point goes there, etc. Since permutations are functions, they can be composed, that is, applied one after another. The permutation fg is f followed by g. Using our convention, fg can be described, as a sequence of 6 numbers, by applying g to the numbers of f. Pretend like the digits of f are animals on a shelf, and move them about according to g. The result is fg. Let's see why this works. Let g start with 6, and let f end in 3. Apply f, and the cat, which is third in line, moves to position 6. Apply g and the cat, which is now in position 6, moves to position 1. Now let's compose the functions and see what we get. Let g act on the digits of f. This moves the number 3, which is in the sixth slot of f, into position 1. Therefore fg starts with 3. Apply fg and the cat moves into position 1. This is the same as f followed by g. There is then an easy way to compose permutations; let each permutation act on the digits of the permutation that came before.

Since permutations are functions, composition is automatically associative. We don't need any tedious algebra to prove it. Apply fg, then h, and it's the same as applying f, then g, then h, which is the same as applying f, and then gh.. In each case the bear passes through f, then g, then h, and winds up in his new slot.

Permutations are not commutative. Let f swap the bear and the cat, and let g swap the cat and the dog. f moves the bear first to position 3, and g then moves it to to position 4. Thus fg moves the bear to position 4. When you apply gf to the bear, g leaves the bear in positon 2, and f moves the bear to position 3.

The identity permutation is 123456. This leaves all the animals in place. It also leaves digits in place, so compose f with the identity permutation and you still have f.

Let f be the permutation 213564. The orbit of an animal is the slots that it moves through as f is applied again and again. Since f swaps 1 and 2, the ape moves to slot 2, then 1, then 2, then 1, and so on. This is its orbit, a path of length 2. The bear has the same orbit. The cat has an orbit of size 1, since it sits in place. Finally, the dog has an orbit of size 3 as it moves to positions 6, 5, and then back to 4. This permutation consists of 3 different cycles, or orbits, having sizes 2, 1, and 3. Cycle notation can be used to describe these orbits. The letter c is used for cycle. Thus $f = c2c1c3$.

If c3 appears twice, we can write $c3^2$, rather than c3c3. Finally, + is used to separate distinct permutations. The cycle decomposition of all the 4-element permutations is described by:

$$6c4 + 8c3c1 + 6c2c1^2 + 3c2^2 + c1^4$$

There are $24 = 4 \times 3 \times 2 \times 1$ permutations on four elements. $6+8+6+3+1 = 24$.

1234	$c1^4$	$(1 \to 1)(2 \to 2)(3 \to 3)(4 \to 4)$
1243	$c2c1^2$	$(1 \to 1)(2 \to 2)(3 \to 4 \to 3)$
1324	$c2c1^2$	$(1 \to 1)(2 \to 3 \to 2)(4 \to 4)$
1342	$c3c1$	$(1 \to 1)(2 \to 3 \to 4 \to 2)$
1423	$c3c1$	$(1 \to 1)(2 \to 4 \to 3 \to 2)$
1432	$c2c1^2$	$(1 \to 1)(2 \to 4 \to 2)(3 \to 3)$
2134	$c2c1^2$	$(1 \to 2 \to 1)(3 \to 3)(4 \to 4)$
2143	$c2^2$	$(1 \to 2 \to 1)(3 \to 4 \to 3)$
2314	$c3c1$	$(1 \to 2 \to 3 \to 1)(4 \to 4)$
2341	$c4$	$(1 \to 2 \to 3 \to 4 \to 1)$
2413	$c4$	$(1 \to 2 \to 4 \to 3 \to 1)$
2431	$c3c1$	$(1 \to 2 \to 4 \to 1)(3 \to 3)$
3124	$c3c1$	$(1 \to 3 \to 2 \to 1)(4 \to 4)$
3142	$c4$	$(1 \to 3 \to 4 \to 2 \to 1)$
3214	$c2c1^2$	$(1 \to 3 \to 1)(2 \to 2)(4 \to 4)$
3241	$c3c1$	$(1 \to 3 \to 4 \to 1)(2 \to 2)$
3412	$c2^2$	$(1 \to 3 \to 1)(2 \to 4 \to 2)$
3421	$c4$	$(1 \to 3 \to 2 \to 4 \to 1)$
4123	$c4$	$(1 \to 4 \to 3 \to 2 \to 1)$
4132	$c3c1$	$(1 \to 4 \to 2 \to 1)(3 \to 3)$
4213	$c3c1$	$(1 \to 4 \to 3 \to 1)(2 \to 2)$
4231	$c2c1^2$	$(1 \to 4 \to 1)(2 \to 2)(3 \to 3)$
4312	$c4$	$(1 \to 4 \to 2 \to 3 \to 1)$
4321	$c2^2$	$(1 \to 4 \to 1)(2 \to 3 \to 2)$

The period of a permutation is the number of times the permutation must be applied to get back to start. A circular shift of 6 elements has a period of 6. The identity permutation has a period of 1. The period is easily calculated, once the permutation is broken up into orbits or cycles. Consider again the permutation 213564. This has a cycle decomposition of $c2c1c3$. The ape and bear participate in a cycle of length 2. The period has to be even if the ape and bear are to return to their original positions. Then there is a cycle of length 3. The period must be a multiple of 3 if the dog eagle and fox are to return to their original positions. The period is then a multiple of 2 and 3, and is 6. In general, the period of a permutation is the least common multiple of its cycles.

Let f permute 9 elements via 345678921.

1 moves to 3
2 moves to 4
3 moves to 5
4 moves to 6
5 moves to 7
6 moves to 8
7 moves to 9
8 moves to 2
9 moves to 1

3 → 5 → 7 → 9→ 1 → 3 is a cycle of 5.
4 → 6 → 8 → 2 → 4 is a cycle of 4.

The cycle decomposition of the permutation f is c5c4, and the period is 20. The c4 cycle is move 8 to 2, because 2 is in position 8; move 2 to 4, because 4 is in position 2; move 4 to 6, because 6 is in position 4; move 6 to 8, because 8 is in position 6. The c5 cycle is move 3 to 5, because 5 is in position 3; move 5 to 7, because 7 is in position 5; move 7 to 9, because 9 is in position 7; move 9 to 1, because 1 is in position 9; move 1 to 3, because 3 is in position 1.

When implementing permutations on a computer, programmers sometimes start numbering at 0, rather than 1. Thus the identity permutation in a group of 6 element permutations is 012345.

Even and Odd Permutations

An adjacent transposition is a permutation that switches two adjacent elements. 123546 Like a common typo that you mihgt make. Note that I define adjacent transposition to be a swap of adjacent elements. This permits the general use of the word transposition for the swap of any two elements, regardless of their position within the permutation.

Any permutation can be built from adjacent transpositions. If you want 3 to wind up at the start, then swap 3 with 2, and then 2 with 1. Next move 5 into position 2 by swapping it with 4, and then with 3, and then 2. Continue all the way to the end, until the permutation looks the way you want. A very simple way to sort a list of words is by making adjacent transpositions until the words are in alphabetical order.

A permutation is even if it is built from an even number of transpositions, and odd if built from an odd number of transpositions. However, is this well-defined? Is there a permutation that can be built with two different chains of transpositions, one even and one odd? The answer is no, so let's prove it.

Given a permutation out of the blue, like 235164, look at each pair of numbers in the permutation and count the pairs that are out of order, wherein the larger number comes first. We will prove that if this count is even, then an even number of transpositions can convert the permutation to 123456, the identity permutation. Therefore, the permutation is even. Also we will have proven than if the count is odd, then an odd number of transpositions can convert the permutation to 123456, the identity permutation. Therefore, the permutation is odd. In our example 235164, 2 comes before 1, 3 comes before 1, 5 comes before both 1 and 4, and 6 comes before 4; that's 5 pairs out of order, a count of 5, which is odd.

We want to prove that if a set of transpositions convert 235164 to 123456, then the number of transpositions in the set is odd.   The permutation 235164 has 5 pairs out of order. Consider any possible transposition. For the permutation 235164, there are 15 possible transpositions.

Apply the transposition 1 <--> 2, to 235164 to get 135264. Now the pairs out of order are (3,2), (5,2), (5,4), and (6,4). We have reduced the number of pairs out of order by 1. The new permutation, 125264 is even.

Apply the transposition 1 <--> 4 to 235164 to get 235461. Now the pairs out of order are (2,1),(3,1) (5,4),(5,1),(4,1), and (6,1). We have increased the number of pairs out of order by 1. The new permutation 235461 is even.

Try it yourself. Apply any transposition to a permutation. Note that the count of pairs out of order will either decrease by 1, or increase by 1. Thus, the count of pairs out of order will always be odd for an odd product of transitions, and even for an even product of transitions.

Let's look at this in great detail.
In general, suppose you apply the transposition (p,q) to a permutation (k1, k2, …kt). Imagine, without loss of generality, that p precedes q in the permutation.

Let x be an element of the permutation that is different than either p or q.

Case 1: x < p and x < q.
Case 1A: x precedes p in the permutation.
(x,p) is in order and (x,q) is in order.
After transposing p and q, it is still true that (x,p) and (x,q) are both in order.
The number of pairs not in order does not change.

Case 1B: p precedes x in the permutation and x precedes q in the permutation.
(p,x) is out of order, and (x,q) is in order.
After transposing p and q, (x,p) is in order, and (q,x) is out of order.
The number of pairs not in order does not change.

Case 1C: q precedes x in the permutation.
(p,x) is out of order, and (q,x) is out of order.
After transposing p and q, it is still true that both (p,x) and (q,x) are out of order.
The number of pairs not in order does not change.

Case 2: x < p and x > q.
Case 2A: x precedes p in the permutation.
(x,p) is in order, and (x,q) is out of order.
After transposing p and q, it is still true that (x,p) is in order, and (x,q) is out of order.
The number of pairs not in order does not change.

Case 2B: p precedes x in the permutation and x precedes q in the permutation.
(p,x) is out of order and (x,q) is out of order.
After transposing p and q, remembering that x < p, and x > q,
(q,x) is in order, and (x,p) is in order.
The number of pairs not in order decreases by 2.
Parity of number of pairs does not change.

Case 2C: q precedes x in the permutation.
(p,x) is out of order, and (q,x) is in order.
After transposing p and q, it is still true that (p,x) is out of order, and (q,x) is in order.

Case 3: p < x and x < q.

Case 3A: x precedes p in the permutation.

(x,q) is in order, and (x,p) is out of order.
After transposing p and q, it is still true that (x,q) is in order, and (x,p) is out of order.
The number of pairs not in order does not change.

Case 3B: p precedes x in the permutation and x precedes q in the permutation.
(q,x) is out of order and (p,x) is in order.
After transposing p and q, remembering that x < q, and x > p,
(p,x) is out of order, and (q,x) is in order.
Parity of number of pairs does not change.

Case 3C: p precedes x in the permutation.
(q,x) is out of order, and (p,x) is in order.
After transposing p and q, it is still true that (q,x) is out of order, and (p,x) is in order.

Case 4: x > p and x > q
Case 4A: x precedes p in the permutation.

(p,x) is in order, and (q,x) is in order.
After transposing p and q, it is still true that (p,x) is in order and (q,x) is in order.
The number of pairs in order does not change.

Case 4B:

Perhaps it will be clearer if we explain in terms of adjacent transpositions.

The identity permutation has 0 pairs out of order, and is even.

An adjacent transposition permutation has one pair of adjacent numbers out of order and is odd. Permuting 215346 to 213546 illustrates an adjacent transposition by swapping the 5 and 3. In the first permutation, 215346, count the numbers preceding 5 that are also greater than 5. There are none. Count the number of numbers preceding 3 that are also greater than 3. There are none. Swapping the adjacent numbers 3 and 5 does not change either of these counts.

In the first permutation, 215346, the 5 is greater than 3. In the second permutation, 213546, the 3 is less than 5. We have reduced the count of out or order numbers by 1. Before there were 3 numbers out of order: 2 before 1, 5 before 3, and 5 before 4. After the permutation, there are 2 numbers out of order: 2 before 1, and 5 before 5. In this case we changed an odd permutation to an even permutation.

In the first permutation, 215346, the 5 is greater than 4, but 3 is less than both 4 and 6. In the second permutation, 213546, 5 is greater than 4, but 3 is less than both 4 and 6. Swapping the 5 and the 3 does not change the count of out of order numbers to the right of the pair being swapped.

Thus, it becomes clear why an adjacent transposition must always change the parity of a permutation. If the pair in the transposition were in order before the permutation, then after the permutation, they are out of order. If the pair in the transposition were out of order before the permutation, then after the permutation, they are in order. If the permutation is even, applying an adjacent transposition must change the parity from even to odd. If the permutation is odd, applying an adjacent transposition must change the parity from odd to even.

Each application of an adjacent transposition flips the parity. Therefore, a chain of adjacent transpositions is even or odd, and the resulting permutation is even or odd. We cannot have an even chain and an odd chain leading to the same permutation, because the even chain creates an even permutation, and the odd chain creates an odd permutation. We have proven that the parity of a permutation is well defined.

There is a homomorphism from the set of all permutations upstairs to even and odd downstairs. The homomorphism is the count function described above. Every permutation maps to even or odd. However, we have to prove it's a homomorphism. Compose functions upstairs, and it's the same as adding even and odd numbers downstairs. This is actually easy once you see the trick. Replace every permutation with a chain of transpositions that builds that permutation. Compose two permutations and you are really putting the two chains together to make one longer chain. Even and odd chains of transpositions add together upstairs, just as even and odd numbers add together downstairs. The homomorphism carries function composition into addition mod 2.

We illustrate with f being a chain of 3 transpositions, and g being a chain of 4 transpositions. Then f+g is a chain of 3+4=7 transpositions.

f is odd, and g is even. f+g is odd. The homomorphism takes even f + odd g into even + odd = odd. Since the permutation f+g is odd, we confirm it is a homomorphism.

Matrices

A matrix is a grid of numbers, so many rows and so many columns. Here is an example that is 3 by 4, 3 rows and 4 columns.

2	1	7	-1
3	5	0	0
-17	-8	3	3

If you program in C, Java, Perl, and most computer languages, the element in the ith row and jth column of a matrix M is denoted $M_{i,j}$, where rows and columns are numbered starting at 0. Mathematicians also use row-order notation, but they usually number rows and columns starting at 1. Thus the upper left entry is $M_{1,1}$. Fortran also starts numbering at 1, but it runs in column order, not row order.

If a variable represents a matrix, it is often capitalized, such as M, A, B, etc. This reminds the reader that a matrix is a larger structure with many numbers inside it. This is a recurring theme. Groups, rings, and even sets are represented with capital letters, because they are larger structures that contain many elements inside them. This sounds simple and clear-cut, but it is actually a blurry distinction. When viewed through ZF set theory, everything is a set, even the number 4. Everything, other than 0, is a set with many elements inside it. So maybe every variable should be a capital letter. On the other hand, a vector in 3 dimensions consists of 3 numbers inside, x y and z, yet it represents a single point in 3 space, so perhaps a small letter v is appropriate. Notation is more of an art than a science. The compromise is to use capital letters for the larger structures in play, and small letters for the elements inside those structures. When there are structures within structures within structures, I will probably use capital letters for the middle layer, and small letters for the elements that are at the bottom; however, if the proofs are broken up into small manageable chunks, this shouldn't happen very often. By analogy, if your software is several hundred lines of code, then it should probably be separated into smaller functions. I hope my proofs are small, simple, and self-contained.

As a general rule, matrices are represented with capital letters, while small letters continue to denote simpler elements like numbers or points in space. Note the next paragraph, where m and n denote the numbers of rows in a matrix M.

A matrix M is m by n, or m×n, if it has m rows and n columns. M is square if m = n.

The main diagonal of a matrix, usually a square matrix, is the set of elements $M_{i,i}$, from upper left to lower right. The identity matrix has ones down the main diagonal and zeros elsewhere.

1	0	0	0
0	1	0	0
0	0	1	0
0	0	0	1

The transpose of a matrix is its reflection through the main diagonal. The transpose of M is written M^T. A matrix is symmetric if it equals its transpose. A symmetric matrix is necessarily square. Here is a symmetric example.

2	1	7	5
1	-8	6	3
7	6	29	0
5	3	0	-4

If two matrices A and B are the same size, then A+B is produced by adding the corresponding entries. In other words, $(A+B)_{i,j} = A_{i,j} + B_{i,j}$.

	1	-4	0
	2	6	x+3
+	5	1	17
	-5	9	2x+5
=	6	-3	17
	-3	15	3x+8

Remember that addition is commutative and associative for real numbers, as described earlier. That means addition is commutative and associative per each entry in the matrix, and thus for matrices as a whole. We have a new set of objects to work with, namely matrices, and a new operator, which is called addition, and addition has the familiar properties: A+B = B+A and (A+B)+C = A+(B+C).

Speaking of objects, matrices are somewhat like a "class" in object-oriented programming. The entries of a matrix can be numbers, but they could be other objects, perhaps other matrices, or quaternions, or polynomials, as shown by the lower right entries in the above example: x+3 + 2x+5 = 3x+8. If addition is defined for the underlying objects, then addition is defined on the matrix. If both multiplication and addition are defined for the underlying objects, then multiplication is defined on the matrix. In fact, many mathematical structures refer to "elements", but often those elements can be other structures, just as an object in C++ can contain other objects. You can have a matrix of polynomials, or you can start with a polynomial in x and substitute a matrix M for x. Thus $x^2 + 4x + 9$ becomes M x M + 4 x M + 9. We'll have many opportunities to put structures within structures, but for now let's return to matrices, and define matrix multiplication, so that M x M makes sense.

Matrix multiplication is not performed element by element; it is more complicated than that. Let's build a matrix C that is A x B. The entry $C_{i,j}$ comes from the ith row of A and the jth column of B. Take the ith row of A and lift it off the page. Rotate it clockwise 90 degrees, so it becomes a column. Line it up with the jth column of B. This creates a double column, a column of pairs of numbers. Multiply these pairs together and add up the products. This operation is called a dot product. The total goes in row i column j of C. Do this for each i and j and build C. Note that the width of A must be equal to the height of B, or multiplication is not defined.

1	0	7
-2	3	3

×

0	2
5	-2
5	x

=1×0+0×5+7×5 $(1 \times 2 - 0 \times 2 + 7\ x)$

-2×0+3×5+3×5 $(- 2 \times 2 - 3 \times 2 + 3\ x)$

=35 $(7\ x + 2)$

30        $(3\ x - 10)$

Matrix multiplication is not commutative. Shown below is A × B and B × A; they are not equal. Even more unusual, A and B are nonzero, yet A × B = 0. This doesn't happen in the real numbers. In the real numbers, if x × y is 0 then either x or y is 0. However, matrices have "zero divisors", blocks that are not zero yet their product is zero.

A × B

0	1
0	0

×

1	0
0	0

=

0	0
0	0

B × A

1	0
0	0

×

0	1
0	0

=

0	1
0	0

Multiplication is distributive over addition. Show this for one entry, and it's true for all the other entries by the same reasoning. Look at the upper left entry of AB + AC. This is the top row of A, $[A_{1,1} \ A_{1,2}]$ dot the first column of B, $[B_{1,2} \ B_{2,1}]$, plus the top row of A, $[A_{1,1} \ A_{1,2}]$, dot the first column of C, $[C_{1,1} \ C_{2,1}]$. Specifically, it is $[A_{1,1} \ A_{1,2}]$ dot $[B_{1,2} \ B_{2,1}]$ plus $[A_{1,1} \ A_{1,2}]$ dot $[C_{1,1} \ C_{2,1}] = A_{1,1} \times B_{1,2} + A_{1,2} \times B_{2,1} + A_{1,1} \times C_{1,1} + A_{1,2} \times C_{2,1}$.

Since multiplication is distributive over addition in the reals,

$A_{1,1} \times B_{1,2} + A_{1,2} \times B_{2,1} + A_{1,1} \times C_{1,1} + A_{1,2} \times C_{2,1} = A_{1,1} (B_{1,2} + C_{1,1}) + A_{1,2} (B_{2,1} + C_{2,1})$

$= [A_{1,1} \quad A_{1,2}]$ dot $[\ (B_{1,2} + C_{1,1}) \quad (B_{2,1} + C_{2,1}) \ ]$

the matrix entry in cell 1,1 of A× (B + C).

In general, calculate the sum of $A_{1,i} \times B_{i,1}$ plus the sum of $A_{1,i} \times C_{i,1}$. Put corresponding terms together to get the sum of $A_{1,i} \times B_{i,1} + A_{1,i} \times C_{i,1}$. Apply the distributive property of numbers and get the sum of $A_{1,i} \times (B_{i,1}+C_{i,1})$. This is, magically, the upper left entry of A× (B+C). Like the commutative and associative properties that came before, matrix multiplication inherits the distributive property from the entries inside.

Multiplication is of course distributive from the right side as well. (A + B) ×C = A × C + B × C.

Matrix multiplication is associative. This also relies on the associative property of the entries inside, but the proof is a bit more complicated. We'll show it for the upper left entry, and the idea carries across. Compare the upper left entry (A × B) × C with A× (B × C). In either case the result is the sum of $A_{1,i} \times B_{i,j} \times C_{j,1}$. I'll let you fill in the algebraic details.

At the end of the day, matrices can be manipulated almost like numbers. You can rearrange a sum of matrices any way you like, because addition is commutative and associative.

Moreover, you can rewrite A× (B + C) × D as A × B × D + A × C × D. The only thing that is different is that A B is not equal to B A. Thus $(A + B)^2 = A^2 + A B + B A + B^2$; don't be tempted to collapse A B + B A into 2 A B.

Linear Functions

Consider 3 dimensional space with the usual x y and z coordinates. A vector v is a point in space with specific values for x y and z, or equivalently, it is the segment from the origin to this point. It has a direction (away from the origin) and a length. Sometimes it is drawn as an arrow pointing outward from the origin. I'll illustrate with the point [2, 3 ,1]. Imagine the point located by starting at the origin, moving 2 units horizontally in the positive x direction, then from there, moving 3 units perpendicularly in the positive y direction, then from there, moving 1 unit upward in the positive z direction. Now instead, imagine a directed line segment going from the origin straight to that point. This is a vector pointing into the first octant, the vector [2, 3, 1]. It's length is sqrt(14).

Vectors provide a geometric intuition for adding points together in space. The following presumes a two dimensional space. Imagine the vector u going from the origin to the point at x = 2, y = 1, and also the vector v, going from the origin to the point at x = 1, y = 2. The origin is the tail of both vectors u and v. The head of vector u is the point [x=2,y=1]. The head of vector v is [x=1,y=2]. Where is the vector u + v? Simply add up the x coordinates, and add up the y coordinates. 2 + 1 = 3, and 1 + 2 = 3. The tail of vector u+v is the origin, and the head of vector u+v is the point[3,3]. Now do the same thing with geometry, rather than algebra. Imagine lifting the vector v, which goes from the origin to the point [x=1,y=2], and moving it parallel so that its tail is now at the head of u. When you do this, you will note that the head of u+v is at [x=3, y=3].

You can also draw the vector v+u by this means. Place the tail of u at the head of v, (i.e. at the point [x=1,y=2]). Draw both vectors u+v and v+u by this method, and complete a parallelogram. The vector from the origin to the point u+v, (or v+u since addition of vectors is commutative), is a diagonal through the parallelogram. This illustrates the general notion that when you place the tail of one vector at the head of another, you are adding the vectors together.

In general, points in space, or vectors in space, can be added together simply by adding their coordinates.

In the real world, we add vectors together to determine the overall velocity or force. Suppose you are traveling east on your bike at 10 miles an hour, and you throw a softball northeast at 10 miles an hour. Draw the implied parallelogram, and the diagonal is the net velocity of the ball, east by northeast at 18.477 miles an hour.

Here is the geometry. Draw an arrow pointing to the right, then at the head of that arrow, draw another arrow of the same length, upward at a 45 degree angle. The horizontal arrow represents traveling east on your bike at 10 miles per hour. The upward slanting arrow, at 45 degrees, represents the path of the thrown softball. The arrow from the origin to the head of the upward slanting arrow represents the net velocity of the ball. The velocity of the ball is then the length of that vector sum. The vector describing the bike velocity is [10,0]. The vector describing the softball velocity, with respect to the bike, is [10 sqrt(½), 10 sqrt(½)] . The sum of the two vectors is [10+10sqrt(½),10sqrt(½)]. By the Pythagorean theorem, which is coming up later in this chapter, The length of that sum is:

sqrt((10 + 10 sqrt(½))^2 + (10 sqrt(½))^2) =

10 × sqrt((1 + sqrt(½))^2 + sqrt(½)^2) =

10 × sqrt((1 + 2 sqrt(½) + ½ + ½) =

10 × sqrt(2 + 2 sqrt(½)) = 18.477.

Let f be a function from one instance of 3 space into another, or from 3 space into itself if you prefer. It is sometimes easier to visualize a function from one space into another, so you aren't stepping on your own feet so to speak. Such a function is linear if f(u+v) = f(u) + f(v). You can add vectors together first, then apply f, or apply f and add the results. This is, wait for it, a homomorphism that respects addition. It also respects scaling, f(c×x) = c×f(x). An entire branch of mathematics, linear algebra, is devoted to these homomorphisms, linear functions from one vector space into another.

The most convenient vectors to use are unit vectors. These are the vectors that point along the x y and z axes, and have length 1. Their coordinates are [1,0,0], [0,1,0], and [0,0,1]. They define the unit cube in space.

Any point p in space is now a linear combination of these three vectors. [8,7,-1] is 8 times the first vector + 7 times the second vector + -1 times the third vector. Other vectors span (cover) space just as well, but they aren't as convenient. For example, let u = [2,1], and let v = [1,2]. These two vectors define a parallelogram, but they also span the entire plane. Every point p in the plane is some real multiple of u plus some real multiple of v - but it's a bit tricky to backtrack and write p as a combination of u and v.

Let's take a closer look at tiling the plane with the vectors u and v. Draw the initial parallelogram on graph paper, having the four sides: [x=0,y=0] to [x=2,y=1], [x=2,y=1] to [x=3,y=3], [x=3,y=3] to [x=1,y=2],and [x=1,y=2] to [x=0,y=0]. This is the "base cell", and I will often refer to the base cell throughout this book. Next, add u to itself, giving the point 2u. In other words, [x=2,y=1] + [x=2,y=1] = [x=4,y=2]. Show this addition on paper by drawing the line segment from [x=2,y=1] to [x=4,y=2]. Next show the addition of v=[x=1,y=2] to 2u=[x=4,y=2] by drawing the line segment from [x=4,y=2] to [x=4+1,y=2+2]. Then show the addition of u=[x=2,y=1] to (u+v)=[x=3,y=3] by drawing the line segment from [x=3,y=3] to [x=3+2,y=3+1]. The end points of the two line segments just drawn are the same, [x=5,y=4]. You have drawn a neighboring parallelogram, sharing a common side with the base cell, from [2,1] to [3,3]. Continue drawing parallelograms in a slanted line on the page, then a row of parallelograms below, then a row of parallelograms above, and so on. For all integers j and k, the point represented by the vector j × u + k × v will be the lower left corner of some parallelogram in the plane. The origin, 0×u + 0×v, is the lower left corner of the base cell. Every point in the plane will either be on or inside one of these parallelograms. This is what we mean when we say we can tile the plane with parallelograms. Furthermore, any point in any parallelogram can be shifted into the base cell by subtracting an integer multiple of u and an integer multiple of v.

Matrix as Function

A permutation is a list of numbers, but these numbers define a function, a map that tells you how to move things about. A matrix is also a function, a linear function from one vector space into another; and each such linear function is represented by a unique matrix. Let's clarify with an example. We multiply the row vector [x, y, z] by a 3 by 3 matrix, M.

[x,   y,   z]

            ×

[1   0   0]

[1   1   0]

[3   7   9]

=[x+y+3z,   y+7z, 9z]

We got that product because [x, y, z] times the column vector [1,1,3] is [x × 1, y × 1, z × 3] and [x, y, z] times the column vector [0,1,7] is [x × 0, y × 1, z × 7], and [x, y, z] times the column vector [0,0,9] is [x × 0, y × 0, z × 9].

Here the x, y and z axes form a coordinate system for 3 dimensional space in the usual way. Let a function f, defined by the matrix M, map the x axis to the x axis. In other words, 1 goes to 1, 2 goes to 2, -37 goes to -37, and so on. Other than squashing the axis down to the origin (which is possible), this is about as simple as it gets. You can realize this by multiplying [x, 0, 0] by M. You get [x, 0, 0] back again. Note however that [0, y, 0] times M is [y, y, 0], and [0, 0, z] times M is [3 z,7 z,9 z].

Next, tilt the y axis over into the first quadrant. Thus 1 (on the y axis) maps to the point [1,1,0], 2 maps to [2,2,0], 9 maps to [9,9,0], and so on. You can realize this by multiplying [0,1,0] by M, [0,2,0] by M, and [0,9,0] by M.

Finally let f(z) be a funky vector out in 3 space, namely [3,7,9]. The unit z vector maps to [3,7,9], twice this vector maps to [6,14,18], and so on. You can realize this by multiplying [0,0,1] by M.

Notice that the images of the axes are the three rows of M. With this in place, f(x,y,z) is [x,y,z] times M, using matrix multiplication. Let's see why this works. Start with a point on the x axis. The y and z coordinates are 0. Represent this point as [x, 0, 0]. Multiply this by M, using matrix multiplication, and get [x, 0, 0]. In other words, we obtain x times the top row of M. Indeed, the x axis maps to itself.

Do the same for [0, y, 0] and extract y times the second row, or [y, y, 0]. Do the same for [0, 0, z] and extract z times the third row, or [3 z,7 z,9 z].

The key here is that f is linear. Let p = [2, 3, 5], a point in space that certainly does not lie on the x y or z axis. f (p) is now f (2, 0, 0) + f (0, 3, 0) + f (0, 0, 5). Matrix multiplication performs this operation for you. The contribution from f (x) is added to the contribution from f (y), is added to the contribution from f (z), as one would expect from a linear function f. Try it with the generic vector [x, y, z]. Multiply this by M and find a new point in 3 space, f (p), that is the same as f (x) + f (y) + f (z). Thus the matrix M establishes the linear function f, and conversely, the images of f along the axes build the matrix M. The correspondence between matrices and linear functions is perfect, one for one.

The next thing to verify is that function composition is the same as matrix multiplication. If a linear function f is represented by a matrix A, and a linear function g is represented by a matrix B, then the combined function fg is represented by A ×B. Illustrate with [x, 0, 0]. Since all the functions are linear, this is enough. x × A extracts x times the top row of A. Multiply this by B to get fg applied to the x axis. This is x times the top row of A, dot each of the three columns of B in turn. Next, multiply A×B, whence the top row becomes the top row of A dot each of the 3 columns of B in turn. Multiply this by [x,0,0] on the left, which scales this row by x. The result is the same. Function composition is the same as matrix multiplication.

A

=

[row 1]

[row 2]

[row 3]

B

=

[Column1   Column 2   Column 3]

A × B

=

[Row 1 dot Column   Row 1 dot Column 2   Row 1 dot Column 3]

[Row 2 dot Column   Row 2 dot Column 2   Row 2 dot Column 3]

[Row 3 dot Column 1   Row 3 dot Column 2   Row 3 dot Column 3]

With this established, there was (perhaps) no need to do all that algebra to prove matrix multiplication is associative. Matrices and linear functions correspond, and function composition is associative, so matrix multiplication is associative. Boom! Then again, matrices can hold entries other than numbers, whence they do not correspond to functions anymore, so maybe it was worth verifying associativity from first principles after all.

Determinant

The determinant of a square matrix can be defined in at least 5 equivalent ways. It is worth exploring this topic in some detail, because the determinant is vital to geometry (formula for volume), calculus (Jacobian), and algebraic number theory (norm/discriminant), to name a few.

In two dimensions, referring to a square matrix with two rows and two columns, the determinant is the product of the upper left times the lower right, minus the product of the upper right times the lower left. This is the area of the parallelogram spanned by the two vectors that form the matrix. (We'll prove this below.) As an example, let two vectors, u = [2,1] and v = [1,2], span a parallelogram in the plane. The area of this parallelogram is the determinant of the matrix whose rows are [2,1] and [1,2]. This is 2×2 - 1×1, or 3. Let's confirm this using high school geometry. This parallelogram is actually a rhombus, so draw the two diagonals, thus cutting the parallelogram into 4 congruent right triangles. One diagonal goes from the point [1,2] to the point [2,1]. The midpoint of the parallelogram is at [1.5,1.5]. The legs of the 4 congruent right triangles are along the diagonals. Each triangle has legs of length sqrt(2)/2 and 3sqrt(2)/2. Each triangle has area (sqrt(2)/2) × (3 sqrt(2)/2)/2 = (2×3/4)/2 = 3/4, and the rhombus has area 3.

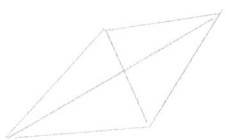

What if u and v both point in the same direction? The parallelogram is flat, and has no area. Assume u is [2,1] as above, but now v is [4,2]. In other words, v is twice as long as u, and points in the same direction. Let u be the first row of M and let v be the second. det(M) = 2×2 - 1×4 = 0. This confirms the fact that the parallelogram has 0 area.

Draw this on graph paper. Connect [0,0] to [2,1] and [4,2], the two sides of the parallelogram. Lift the first side up and place it on the end of the second side, from [4,2] to [6,3]. This is the third side of the shape, parallel to the first. Then draw the roof, from [2,1] to [6,3]. You tried to draw a parallelogram, but it is actually a flat line, as though it had been run over by a truck.

Now move to 3 vectors in 3 space. These span a parallelotope, which is like a pushed over box (the way a parallelogram is a pushed over rectangle). The determinant gives the volume, which would be hard to figure out from geometric principles alone. Call the vectors [a,b,c], [d,e,f], and [g,h,i], and arrange them in a 3 by 3 matrix.

[a b c]

[d e f]

[g h i]

The determinant of a three by three matrix is the sum of 3 products minus the sum of 3 more products. Draw a line through the main diagonal and multiply these three elements together; this is the first product: $a \times e \times i$. Shift the line to the right, and let it wrap around the matrix at the bottom, so that it passes through the top row middle entry, the middle row right, and the bottom row left, then multiply these together: $b \times f \times g$. Finally shift the line to the right again, so that it passes through the top row-right, the middle row-left, and the bottom row-middle, giving the third product: $c \times d \times h$. These are the positive products: aei + bfg + cdh. Three more products will be subtracted from this total. Draw a line from the upper row right to the lower row left. This is sometimes called the antidiagonal, because it runs opposite to the main diagonal. This passes through three entries; multiply them together and subtract from the previous total: $-c \times e \times g$. Shift the line to the left, wrapping around the matrix at the bottom, so that it passes through the top row middle, the middle row left, and the bottom row right. Multiply these together and subtract: $-b \times d \times i$. Finally, multiply together the top row left, the middle row right, and the bottom row middle: $-a \times f \times h$. Just to review, the lines that slant parallel to the main diagonal are the positive products, and the lines that slant perpendicular to the main diagonal are the negative products. (This was also the case for the 2 by 2 matrix.)

+ aei + bfg + cdh - ceg - bdi - afh

If a 3 by 3 matrix has rows [1,2,3], [4,5,6], and [7,8,9], then its determinant is $1 \times 5 \times 9 + 2 \times 6 \times 7 + 3 \times 4 \times 8 - 3 \times 5 \times 7 - 2 \times 4 \times 9 - 1 \times 6 \times 8 = 45 + 84 + 96 - 105 - 72 - 48 = (96-105) + (84-72) + (45-48) = (-9 + 12 - 3) = 0$.

Here is another example 3 by 3 matrix; verify that its determinant is 2.

2 2 1

1 1 1

3 1 4

$2 \times 1 \times 4 + 2 \times 1 \times 3 + 1 \times 1 \times 1 - 2 \times 1 \times 1 - 2 \times 1 \times 4 - 1 \times 1 \times 3$

= 8+6+1-2-8-3 = 2.

The volume of the parallelotope spanned by these vectors is 2. But what if the last vector (the bottom row) was 3,3,4? The determinant comes out 0. This makes sense, because the box is more than pushed over, it is flat. All three vectors live in the plane x = y. The box has no volume because it has been squashed into Flatland, and the determinant agrees.

Before we prove all these assertions about area and volume, Step back to one dimension, which is just a line. The vector from 0 to 9 has length 9. Length is volume when you are stuck in one dimension, so the volume spanned by this single vector is 9. At the same time, the 1 by 1 matrix containing this vector is the single entry 9. The determinant of a 1 by 1 matrix is its entry, so the determinant is 9. In one dimension it is clear that determinant equals volume. Well almost. The vector -9 still has length 9, but its determinant is -9. So technically we must take the absolute value of the determinant to get volume, though there will be times when the sign of the determinant is significant.

It would be nice to tackle the determinant, and perhaps volume, in n dimensions, all in one go, but so far I have only defined the determinant for 1 2 and 3 dimensions. How about 4 vectors in 4 space, building a 4 by 4 matrix? Volume is hard to picture in higher dimensions, but it is still well defined, with the unit cube having volume 1. Volume and determinant still agree; but the formula for determinant gets much more complicated. You can't just draw slanted lines in two directions and add/subtract products. So let's define determinant recursively for all dimensions.

1.   As described above, the determinant of a 1 by 1 matrix is the single entry in that matrix. Given a larger matrix, delete the first row and the jth column. This leaves n-1 rows and n-1 columns. We set up a mathematical induction by assuming we can compute the determinant of this smaller matrix. This is called a subdeterminant. Multiply this by -1 if j is even. This is called a cofactor. Finally multiply the cofactor by the entry in row 1 column j. Add up all these products as j runs from 1 to n to obtain det(M). Call this recursive procedure det1(M). In the following 5 by 5 example, I have excised the first row and fourth column , and put stars in as placeholders, except for row 1 column 4, which I left as 7 from the original matrix. This leaves a 4 by 4 matrix with subdeterminant -270, a cofactor of 270 (since j is even), and a contribution of 7×270 or 1890 to det1(M).

*	*	*	7	*
2	3	-1	*	-4
0	9	6	*	1
0	0	3	*	3
0	0	0	*	-5

2.   If the matrix is 2 by 2, with entries a and b on top and c and d below, det1(M) is thus a times the determinant of d, - b times the determinant of c, which is ad-bc, which is the formula given earlier. So far so good.

3.   Verify, in the same way, that det1(M) for a 3 by 3 matrix gives the 6 term formula that was described with slanted lines traveling parallel to and perpendicular to the main diagonal. We're still on track.

[a b c]

[d e f]

[g h i]

Then det1 = a× (e × i – f × h) – b× (d × i – f × g) + c× (d × h – e × g)

= a×e×i + b×f×g + c×d×h – a×f×h – b×d×i – c×e×g

4.   Develop the formula for det1(M) for a 4 by 4 matrix if you like, but it has 24 products, and gets kinda messy. The formula for an n by n matrix has n factorial terms and becomes quite unmanageable. Pause for a moment and prove this by induction on n.

The determinant of a one by one matrix has a single term. The determinant of a two by two matrix has two terms. The number of terms in the determinant of a three by three matrix is the number of columns of row 1 times the number of terms in the determinant of a two by two matrix. That is, the number of terms in the determinant of a three by three matrix is 3 times the number of terms in the determinant of a two by two matrix. $3 \times 2 = 6$

Now you see the sequence. The determinant of a four by four matrix has 4×6 terms.

The determinant of a five by five matrix has $5 \times (4 \times 6)$ terms. And so forth.

Note that the det1 definition of determinant is defined in terms of recursively taking sub-determinants while marching down the rows.

5.   The second definition, det2(M), is based on permutations. Choose any entry from the first row and mark that column as "used". Then choose an entry from the second row, avoiding the used column. Mark that column used, and select an unused entry in the third row, and so on. Continue down to row n. The columns selected, for rows 1 through n, define a permutation. Each column is listed once and only once, and there are n! ways to do this. Calculate the product of entries for each permutation, and negate it if the permutation is odd. Finally add up all these products to get det2(M).

Consider again the 2 by 2 matrix with "a" and "b" on top and "c" and "d" below. We can select "a" from the first row, which forces "d" from the second. The columns selected are 1 and 2, in that order, and that is an even permutation. Thus the product is ad. The only other possibility is bc. This permutation is 2 1, which is odd, so it becomes -bc. The determinant is ad-bc, which agrees with det1(M).

If M is 3 by 3, there are 6 permutations, 3 even and 3 odd. This leads to 6 products, 3 of them negated. Show that this too agrees with det1(M).

Proceed by induction through higher dimensions. Given an n by n matrix, delete the first row and jth column, and write the subdeterminant as a sum of permutation products. If n = 6, for example, the subdeterminant is 5 by 5, and one of the permutations is 21543. What would be the product for this permutation? Find out whether this permutation is odd or even. This is accomplished with two swaps, (2,1) and (3,5). Therefore this permutation is even. Therefore the product of the corresponding entries is positive.

Now add 1 to each column number that is j or larger. We are sliding them over to make room for column j. If j = 4 this changes our example to 21653. Then slide j in position, as if it had been there all along, giving 216453. This permutation still consists of two swaps, like the original, and is still even, but j is in the wrong place; j is suppose to be first in line. After all, det1 says to multiply the subdeterminant by the entry in row 1 column j, so j comes first. A series of j-1 transpositions moves j to the head of the line. This changes 216453 into 421653.

Have we changed the parity of the permutation by pulling j to the front? Only if an odd number of transpositions are needed to move j from its home in position j to the front in position 1. This happens only when j is even. In the same way, det1(M) tells us to negate the product precisely when j is even. The permutation products implied by det1 and the permutation products of det2 are the same, and the two formulas agree.

Here is a sample 4 by 4 matrix with entries 1 through 16. This is followed by a list of the 24 permutations, with the sign (+ for even parity and - for odd parity), and the permutation product. Add these up to get a determinant of 0.

1	2	3	4
5	6	7	8
9	10	11	12
13	14	15	16

+[1,2,3,4] → +1×6×11×16

-[1,2,4,3] → -1×6×12×15

-[1,3,2,4] → -1×7×10×16

+[1,3,4,2] → +1×7×12×14

+[1,4,2,3] → +1×8×10×15

-[1,4,3,2] → -1×8×11×14

-[2,1,3,4] → -2×5×11×16

+[2,1,4,3] → +2×5×12×15

+[2,3,1,4] → +2×7×9×16

-[2,3,4,1] → -2×7×12×13

-[2,4,1,3] → -2×8×9×15

+[2,4,3,1] → +2×8×11×13

+[3,1,2,4] → +3×5×10×16

-[3,1,4,2] → -3×5×12×14

-[3,2,1,4] → -3×6×9×16

+[3,2,4,1] → +3×6×12×13

+[3,4,1,2] → +3×8×9×14

-[3,4,2,1] → -3×8×10×13

-[4,1,2,3] → -4×5×10×15

+[4,1,3,2] → +4×5×11×14

+[4,2,1,3] → +4×6×9×15

-[4,2,3,1] → -4×6×11×13

-[4,3,1,2] → -4×7×9×14

+[4,3,2,1] → +4×7×10×13

Why is the determinant of this 4 by 4 matrix equal to 0? If you add the first column to the fourth column, then subtract the second and third columns, the result is a column of all zeros. We'll see below that the determinant is 0 exactly when some linear combination of rows, or columns, comes out all zeros.

6.   Let M be any matrix and let S be M with two adjacent rows swapped. Compare det2(M) and det2(S). Let p be a permutation product in det(M). Let p use column 7 for the first row and column 9 for the second. Let q be the corresponding permutation product on S, with 7 and 9 swapped. Now p and q multiply exactly the same entries together, but in a different order. Since multiplication is commutative, order doesn't matter. The two products are equal. Actually there is a catch; we swapped 7 and 9, and that changed the parity of the permutation. Thus the corresponding products are opposite, and det(S) = -det(M).

Every time we swap adjacent rows, or we interchange any two rows, or apply an odd permutation to the rows of a matrix, the determinant is negated.

Note that det2 method of calculating the determinant uses permutations of the columns as you march down the rows.

Using this swap principle, we can generalize det1(M). The new formula, det3(M), is recursive, like det1(M), but you can designate any row, not just the first row. Delete the ith row and jth column, as j runs from 1 to n. Compute the subdeterminant, and negate it when i+j is odd (giving the cofactors for that row), and multiply by Mi,j. Does this give the right answer? It does when i = 1. Note that i+j is odd precisely when j is even (since i = 1). Thus det1 and det3 are the same when i = 1. Proceed by induction on i.

Swap row i with row i+1 and apply det3 to row i+1, remembering that det3 was correct when the same row was in position i. The subdeterminants are the same, and the products are the same, except i+j is odd when it used to be even, and even when it use to be odd, since i has increased by 1. The determinant has been negated, at least according to det3. However, we know the determinant of a matrix is negated when two rows are swapped, so det3 agrees with det1 and det2.

Note that the det3 definition of determinant is defined in terms of recursively taking sub-determinants while marching down a permutation of the rows.

7.   Let M be any matrix and let S be its transpose. Compare det2(M) and det2(S). Let p be a permutation and let q be its inverse. For example p = 1 3 5 7 2 4 6. Following the notation used earlier in the permutation chapter, we see that p moves 1 to position 1, 2 to position 3, 3 to position 5, 4 to position 7, 5 to position 2, 6 to position 4, and 7 to position 6.

Thus to undo permutation p, we would move 1 to position 1, move 2 to position 5, move 3 to position 2, move 4 to position 6, move 5 to position 3, move 6 to position 7, and move 7 to positon 4.

The inverse of p is q = 1 5 2 6 3 7 4.

Confirm by seeing that permutation p followed by permutation q is the identity permutation that leaves everything in place.

We apply this to the determinants of M and S.

The product defined by p includes the entry in the third row and second column of M, and the product defined by q includes the entry in the second row and the third column of S. Since S is the transpose of M, these are precisely the same. The two permutation products are the same. Note also that p and q have the same parity. This is because pq gives the identity permutation, where everything is fixed in place. For example, pq = 1234567. The result of p followed by q is even, so either p and q are even, or p and q are odd. The implied products are both positive or both negative. Put all this together and det2(M) = det2(S).

Det4 is defined recursively, just like det3. Select any column j and remove it. Travel down this column and remove each row i in turn. Compute the sub-determinant, and negate it when (i+j) is odd, giving the cofactors for column j. Multiply by the entries in column j and add up the results to get det4(M).

DET4 = M1j × subdet(1,j) + M2j × subdet(2,j) + …

The proof of equivalence is easy. Take the transpose of everything, which does not change det(M). We are now marching across a row, rather than down a column. This is det3; Confirm that formula for det3 and det4 agree.

Note that the det4 definition of determinant is defined in terms of recursively taking sub-determinants while marching down a permutation of the columns.

8.   This definition, det5(M), assumes M is upper triangular. In other words, everything below the main diagonal is 0. Use det4 and march down the first column. All the entries, other than the first entry, are 0, and all the products, other than the first one, are 0. We only have the upper left entry times the subdeterminant of M sans its top row and left column.

Yet this submatrix is also upper triangular, and its determinant can be found in exactly the same way. Take the upper left entry and multiply by the next subdeterminant. Continue all the way down to the lower right, and det5(M) is the product of the entries on the main diagonal.

The same holds for a lower triangular matrix, wherein everything above the main diagonal is 0.

2	1	7	-1
0	5	3	0
0	0	-8	3
0	0	0	-2

2	*	*	*
*	5	3	0
*	0	-8	3
*	0	0	-2

Gaussian Elimination

Recall the definitions of det1, det2, det3, and det4.

Note that the det1 definition of determinant is defined in terms of recursively taking sub-determinants while marching down the rows.

Note that det2 method of calculating the determinant uses permutations of the columns as you march down the rows.

Note that the det3 definition of determinant is defined in terms of recursively taking sub-determinants while marching down a permutation of the rows.

Note that the det4 definition of determinant is defined in terms of recursively taking sub-determinants while marching down a permutation of the columns.

These definitions of determinant are interesting, but completely impractical. You can program a computer to compute det2(M), but the number of permutation products is n factorial.

When n = 100, (and yes we do sometimes work with matrices this large and larger), 100! is just a huge number; the sun would grow cold before the computer was finished. And det1, det3, and det4 aren't any better. The last formula, det5, is plenty fast enough, but it only applies to a special kind of matrix, namely upper triangular. As Gauss discovered, there is a way to turn any matrix into an upper triangular matrix, and that is the key. First, we need a few lemmas.

Scale a row of M by x, and you multiply the determinant by x. For example, multiply the third row of M by 7, and this multiplies the determinant by 7. This is clear if you use det3, and march along the third row. Each entry in this row, and hence each product, is multiplied by 7, hence det(M) is multiplied by 7.

The same proof holds if you scale a column; just use det4(M).

If a row or column of M is 0 the determinant is 0. This is clear from det3 or det4.

If the third row of M is the sum of two vectors u and v, then det(M) is the sum of the two determinants using u alone and v alone. What does this mean? Assume you have three matrices that are identical except for the third row. In matrix 1, the third row is a vector u. In matrix 2, the third row is a vector v. In matrix 3, the third row is the vector which is the sum of vector u and vector v. The determinant of matrix 3 is then the sum of the determinants of matrix 1 and matrix 2. This is easy to prove using det3; just march along the third row and use uj+vj, or ujalone, or vj alone.

This is actually an application of the distributive law: C(A+B) = C A + C B.

Consider the special case when the second row is k times the first, and consider det2(M). Group the permutations into pairs, where the first two numbers switch places. For instance, one permutation might begin with 3 5 while its counterpart begins with 5 3. Let the first row of M have x in column 3 and y in column 5, which means the second row of M must have kx in column 3 and ky in column 5. Both products now start out with kxy. However, since 3 and 5 swap, one permutation is even and the other is odd. One product is positive and the other is negative. They cancel each other out; and this happens for each pair of permutation products. Therefore det(M) is 0. This happens whenever any row is k times another row, or whenever any column is k times another column. Here is a 3 by 3 example where the second row is twice the first.

1	2	3
2	4	6
7	8	9

The determinant is $1{\times}4{\times}9{-}2{\times}2{\times}9 + 3{\times}2{\times}8{-}1{\times}6{\times}8 + 2{\times}6{\times}7{-}3{\times}4{\times}7$. Note that the pairs of terms drop to 0.

Now we can see the magic. Let the first row of M be a vector u, and let the second row of M be a vector v. Add k times v to u in the first row of M. The new determinant, based on u+kv, is the determinant with u alone plus the determinant with kv alone. But when kv is the first row of M it is k times the second row, and that has a determinant of 0. Therefore the determinant is the same whether we use u or u+kv. You can add k times any row of M to any other row, and the determinant does not change.

To visualize this, imagine that there are again three matrices. Matrix 1 has the original first row u. Matrix 2 has the first row equal to kv, i.e. k times the second row, and Matrix 3 has the first row equal to u+kv. As before det(M3) = det(M1) + det(M2); but M2, because its first row is k times the second row, has a determinant equal to zero. Hence det(M3) = det(M1).

Add any multiple of any row to any other row, and it does not change the determinant. The same goes for columns of course.

Now for Gaussian elimination. Start with any matrix M. If the first column is 0 then move on to the second column, as described in the next paragraph. Otherwise find the first entry in column 1 that is not 0. If it is not at the top then swap the two rows so that it is at the top. (If you have to swap rows then remember -1, because you just multiplied det(M) by -1.) Say the upper left entry is x. Proceed down the first column and consider each row in turn. If y is in row 2 column 1, then subtract y/x times the first row from the second. This does not change the determinant, and it turns y into 0. Continue down to the bottom, and the first column becomes 0 except for x at the top.

Consider the second column, ignoring the entry at the top. In other words, the second column effectively begins with position 2. If this number in positon 2 is 0, then move on to the third column, as described in the next paragraph. Otherwise find the first entry in this column that is not 0. If it is not in row 2 then swap the two rows so that it is in row 2. (If you have to swap rows then remember the multiplier, -1, because you just multiplied det(M) by -1.) Say M(2,2) = x. Proceed down the second column and consider each row in turn. If y is in row 5 column 2, then subtract y/x times the second row from the fifth row. This does not change the determinant, and it turns y into 0. Continue down to the bottom, and the second column becomes 0 below position (2,2.

Do this for each column in turn, and M becomes upper triangular. This process is called Gaussian elimination, because we have eliminated the entries below the main diagonal. Now apply det5, and multiply by -1 if necessary. This produces det(M). Here is a 3 by 3 example of gaussian elimination:

4	10	7
3	8	4
1	11	1

Eliminate the entries in the first column by subtracting suitable multiples of the first row from the second and third rows.

4	10	7
3 - ¾×4	8 - ¾×10	4 - ¾×7
1 - ¼×4	11 - ¼×10	1 - ¼×7

4	10	7
0	1/2	-5/4
0	17/2	-3/4

Next eliminate the 17/2 in row 3 and column 2 by subtracting a suitable multiple of row 2.

4	10	7
0	1/2	-5/4
0	17/2 - 17×(1/2) =	-3/4 - 17×(-5/4)

4	10	7
0	1/2	-5/4
0	0	82/4

Multiply the diagonal entries to get a determinant of 41, which agrees with det2(M) having 3 positive products and 3 negative products.

Notice that we had to go from integers to fractions to perform Gaussian elimination. This is usually not a problem. If there are no zero divisors then we're ok. However, there are some rings where Gaussian elimination does not work. The extension into fractions is not feasible. Consider the integers mod 12, and the following matrix.

3	8
2	9

Try to clear out the first column, as outlined above. You can't divide 2 by 3, because the multiples of 3 are 0 3 6 and 9. And swapping rows doesn't help. You can't divide 3 by 2 because all the multiples of 2 are even. Swapping columns doesn't help either. So you're kinda stuck. M has a valid determinant, in this case 11, but you need another procedure to find it. Of course this matrix is rather pathological. Gaussian elimination works most of the time - whenever division is well defined, which it usually is.

Elementary Row Operations

Gaussian elimination relies on elementary row operations. Here they are, along with their effect on the determinant.

1.   Swap two rows: negates the determinant.

2.   Add k times one row to another: leaves the determinant unchanged.

3.   Multiply a row by x: multiplies the determinant by x.

These operations can be accomplished by matrix multiplication. The matrices that affect these changes are called elementary matrices. Below are 3 matrices that correspond to the 3 elementary row operations. When multiplied by M on the right, they swap the second and third rows, add 7 times the fourth row to the first row, and scale the fourth row by 9, respectively.

1	0	0	0
0	0	1	0
0	1	0	0
0	0	0	1

1	0	0	7
0	1	0	0
0	0	1	0
0	0	0	1

1	0	0	0
0	1	0	0
0	0	1	0
0	0	0	9

The determinants of these matrices are -1, 1, and 9, which is precisely the effect these matrices have on the determinant of M. This will become very important in the next section.

There are elementary column operations, just as there are elementary row operations. You simply act on columns, rather than rows. Each of these can be implemented by an elementary matrix E; just multiply M×E, rather than E×M. For example, if E is the first matrix shown above, then M×E swaps columns 2 and 3. This multiplies det(M) by -1, which happens to be the determinant of E.

Sometimes a diagonal matrix is considered elementary, scaling all the rows at once. The following matrix scales the rows of M by 2, 3, 5, and 9. Or it scales the columns if you put E on the right. The determinant of M is multiplied by 2×3×5×9, which happens to be the determinant of E.

2	0	0	0
0	3	0	0
0	0	5	0
0	0	0	9

The Determinant of the Product

We now have enough machinery in place to build a very important homomorphism from matrices down to the underlying elements. The homomorphism is the determinant, which carries a matrix upstairs to a number downstairs. To be a true homomorphism, det(AB) has to equal det(A)×det(B). Let's prove this using elementary row operations.

If E is an elementary matrix and M is any matrix, det(M×E) = det(M)×det(E), and det(E×M) = det(E)×det(M). The theorem is true as long as one of the two matrices is elementary. In fact, the theorem is true for any sequence of elementary matrices.

Given two matrices A and B, turn each into a sequence of elementary matrices, and the determinant of the very long product is the product of the determinants. Everything is associative, so clump the elementary matrices back together, according to A and B, and det(AB) = det(A)×det(B). The key is turning A and B into elementary matrices.

Gaussian elimination takes us halfway there. Turn A into an upper triangular matrix T, then reverse these steps, so that A = E1E2E3…T, where each Ei is an elementary matrix implementing an elementary row operation. That's half the battle. As long as each entry in the main diagonal of T is nonzero, then Gaussian elimination can be performed again, this time clearing the entries above the main diagonal. Subtract multiples of the bottom row of T from the upper rows, thus clearing the rightmost column of T above $T_{n,n}$. Use more elementary matrices to clear the next column, and the next, until T becomes a diagonal matrix D. T has been "diagonalized". Reverse these steps, so that T is D multiplied by another product of elementary matrices on the left. Put this all together to get:

A = E1 E2 E3 … F1 F2 F3 … D

Now A is a product of elementary matrices. Do the same for B, and the theorem is complete.

But there is a catch. What if T has zeros on its main diagonal? This happens iff det(A) = 0. Gaussian elimination always pushes the nonzero rows up to the top, hence the bottom row of T is entirely zero. The bottom row of T×B is also 0, which means det(T×B) = 0. This agrees with det(T)×det(B), which is 0 times det(B), or 0. Bring in the elementary matrices on the left and det(AB) = 0.

On the other side, let det(B) = 0. Perform Gaussian elimination by columns, on the right, making B a lower triangular matrix T times some elementary matrices on the right. If det(B) = 0 then the right most column of T is 0. The right most column of A×T is 0, and det(A×T) = 0. Bring in the elementary matrices on the right and det(AB) = 0. That completes the proof. Determinant is a homomorphism from matrices to numbers that respects multiplication.

Actually there is another caveat. What if A cannot be diagonalized? Recall our 2 by 2 matrix, mod 12, that would not yield to Gaussian elimination, because division was not well defined. If A cannot be written as a product of elementary matrices, what do we do then? As it turns out, it is possible to parlay the product rule from the complex numbers (where division is always well defined) to any commutative ring, including mod 12. Consider 4 by 4 matrices. Let the first matrix consist of 16 variables, just letters if you will, and let the second matrix consist of 16 more variables. Express det(AB) - det(A) ×det(B) as a polynomial in 32 variables. Replace these 32 variables with any 32 complex numbers and the polynomial evaluates to 0. By nullstellensatz, a theorem in algebraic geometry that is many chapters ahead, the polynomial is 0. The relationship holds at an algebraic level, and is valid for anything we care to put in our matrices.

Determinant is more than a homomorphism; it is an epimorphism. Given x, build a matrix M that is the identity matrix, except for the upper left entry, which is x. Now det(M) = x. For every x there is some matrix with determinant x, and that makes determinant an epimorphism.

x	0	0	0
0	1	0	0
0	0	1	0
0	0	0	1

[Here]

Matrix Identities and Inverses

The additive identity for matrices is the zero matrix, wherein every entry is zero. 0+M = M, and M+0 = M.

Every matrix M has an additive inverse, denoted -M, wherein each entry is negated. M + -M = 0.

The identity matrix, described earlier, and reproduced below, is the multiplicative identity. It is sometimes denoted I, or simply 1. Verify that $1 \times M = M \times 1 = M$.

1	0	0	0
0	1	0	0
0	0	1	0
0	0	0	1

The determinant of the identity matrix is 1. Let V be the inverse of M, so that $M \times V = 1$. Take determinants, and $\det(M) \times \det(V) = 1$. If $\det(M) = 0$, then this is impossible. A matrix with 0 determinant has no multiplicative inverse. This is called a singular matrix.

Conversely, assume M has a nonzero determinant. Write M as a string of elementary matrices times a diagonal matrix, as we did above. Each of these matrices can be easily inverted. Here are the elementary matrices again, with their inverses below.

[1	0	0	0
0	0	1	0
0	1	0	0
0	0	0	1]
[1	0	0	7
0	1	0	0
0	0	1	0
0	0	0	1]
[1	0	0	0
0	1	0	0
0	0	1	0
0	0	0	9]

[1   0   0   0

0   0   1   0

0   1   0   0

0   0   0   1]

[1   0   0   -7

0   1   0   0

0   0   1   0

0   0   0   1]

[1   0   0   0

0   1   0   0

0   0   1   0

0   0   0   1/9]

Having written M as a string of elementary matrices, reverse the string and invert each matrix. Multiply these two strings together and get 1. Matrices cancel each other, pair by pair, like matter and antimatter. Invoke the magic of associativity, and clump the first string of elementary matrices together under M, and then the reverse string of inverse elementary matrices under V. Sure enough, M×V = 1, and M is invertible. M is a nonsingular matrix.

Again, this proof is not entirely satisfying, since the following matrix, mod 12, does not yield to Gaussian elimination; yet it is invertible.

3   8

2   9

If the determinant of this matrix was 2, M would not be invertible. You can't multiply 2 by anything to get 1 mod 12. However, this determinant is 3×9 - 2×8 = 11, and 11×11 = 1 mod 12, so there might be a way.

Let d = det(M), and assume d is invertible. Let V be the transpose of the cofactors of M. Remember that the cofactors are the subdeterminants, times 1 or -1 in a checkerboard pattern. This was described in det1(M), and generalized in det3(M). The transpose of the cofactors is called the adjoint. See below for our 2 by 2 example.

Cofactors

9	-2
-8	3

V

9	-8
-2	3

Multiply M by its adjoint and see what happens.

3	8
2	9

×

9	-8
-2	3

=

11	0
0	11

The answer is the determinant running down the main diagonal. Why is this so? When row i lines up with column i, the cofactors of M are multiplied by the entries of M on row i, and the result is d, according to the definition of det3 applied to row i. Thus M×V has the determinant d running down the main diagonal. For row i and column j, j different from i, the dot product is the determinant of a new matrix M', which is the same as M except row j is replaced with row i. Since 2 rows are the same, this determinant is 0. The off diagonal entries are all 0. M×V is a diagonal matrix with each entry set to d. Divide the rows of V by d, which is an elementary operation, and find the inverse of M. In other words, a scaled version of the adjoint of M is the inverse of M, And this inverse works on the left as well as the right. M is invertible iff its determinant is invertible.

Apply this procedure to 2 and 3 dimensions. The generic matrix is on the left, and its inverse is on the right. In each case the determinant is y. The inverse is the adjoint divided by y.

$$\begin{matrix} a & b \\ c & d \end{matrix}$$

$$\begin{matrix} d & -b \\ -c & a \end{matrix}$$

/ y

$$\begin{matrix} a & b & c \\ d & e & f \\ g & h & i \end{matrix}$$

$$\begin{matrix} ei\text{-}fh & ch\text{-}bi & bf\text{-}ce \\ fg\text{-}di & ai\text{-}cg & cd\text{-}af \\ dh\text{-}eg & bg\text{-}ah & ae\text{-}bd \end{matrix}$$

/ y

Suppose M has some other left inverse U, which is different from its right inverse V. Start with UM = 1 and multiply by V on the right, giving UMV = V. More associative magic, and (UM)V is the same as U(MV). Thus U×1 = V, and U = V. There is but one inverse V, and it works on either side.

In summary, if U is a left inverse of M and V is a right inverse of M, then we write U = U×1 = U× (MV) = (U×M) ×V = 1×V = V

This is a general theorem from group theory that is not restricted to matrices. If something is invertible from both sides, the two inverses are the same. Within a ring, a 2-sided invertible element is called a unit.

The product of two units is another unit. Thinking of matrices again, let A and B be units. Let A have inverse U, and let B have inverse V. Verify that VU is the inverse of AB, on either side, making AB a unit.

If u is a unit it cannot be a zero divisor, from either side. They are mutually exclusive. Suppose ux = 0, and multiply by v on the left. Now vux = 0, and since vu = 1, x = 0.

In the integers, the only units are 1 and -1. In the reals, everything other than 0 is a unit.

x and y are associates if there is a unit u such that $xu = y$. In the integers, 5 and -5 are associates.

Multiply x by all the units to get all the associates of x. If z is an associate of y, and y is an associate of x, then $z = (xu)v = x(uv)$, and z is an associate of x. If y is an associate of x then multiply by the inverse of u on the right and x is an associate of y. Finally, $x \times 1 = x$, so x is an associate of itself. Associates clump together into well-defined sets, known as equivalence classes. Other than 0, these sets all have the same size, according to the number of units. In the integers, n and -n are associates, according to 1 and -1.

Return to the 2 by 2 matrices, where multiplication does not commute. Let X be 0 except for 1 in the upper left, and let Y be 0 except for 1 in the lower left. An elementary matrix E swaps the two rows, so that $E \times X = Y$. Thus X and Y are associates from the left. However, there is no invertible matrix, or any matrix for that matter, that turns X into Y from the right. The bottom row of X times anything is 0. Thus X and Y are not right associates.

Orthogonal

You probably saw the law of cosines in high school geometry, but let's review it here. It is a generalization of the Pythagorean Theorem. While we're at it, let's prove the Pythagorean Theorem itself. It's the loveliest and simplest proofs that you will ever see.

If the legs of a right triangle have lengths a and b, and the hypotenuse has length c, the lengths satisfy $a^2 + b^2 = c^2$. For example, set $a = 33$ and $b = 56$, and the hypotenuse has length 65.

Given a right triangle with sides a b and c, draw a square a+b units on a side. Place a copy of the right triangle in each of the four corners of the square. Each triangle points to the next one, like a snake chasing its tail.

Now the bottom of the square, a+b in length, is covered by the a leg of one triangle and the b leg of the next, and similarly for the other three sides. The region enclosed by the four triangles is a square, c units on a side. This inner square is tilted relative to the outer square, but it is still a square, having area c^2. The four triangles have a combined area of 2ab, and the outer square has area $(a+b)^2$. Put this all together and derive $a^2 + b^2 = c^2$. The area of the 4 triangles plus the area of the inside square is equal to the area of the outside square.

$4(\frac{1}{2})(ab) + c^2 = (a+b)^2$

$2ab + c^2 = a^2 + 2ab + b^2$

$c^2 = a^2 + b^2$

Now for the law of cosines. Let a triangle have sides with lengths a b and c, such that the first two segments meet at an angle of t. The law of cosines states:

$a^2 + b^2 = c^2 + 2ab \times \cos(t)$

If t is 90 degrees, its cosine is 0, giving the Pythagorean theorem once again.

Let b be the base, with a and c meeting at the apex. We stipulate that b and a meet at the origin, at angle t. Drop a perpendicular from the apex, splitting b into lengths p and q. Thus p+q = b. Note that p or q could be negative, if the apex leans off to the left or right of the base.

Use the Pythagorean theorem twice to get:

$a^2 - p^2 = c^2 - q^2 = d^2$, where d is the length of the perpendicular from the apex to b.

Replace q with b-p, and p with a×cos(t). This will produce the desired formula.

$a^2 - p^2 = c^2 - q^2$

$a^2 - (a×cos(t))^2 = c^2 - (b - (a×cos(t)))^2$

$a^2 - (a×cos(t))^2 = c^2 - (b^2 - 2ab×cos(t) + (a×cos(t))^2)$

$a^2 - (a×cos(t))^2 = c^2 - b^2 + 2ab×cos(t) - (a×cos(t))^2$

$a^2 = c^2 - b^2 + 2ab×cos(t)$

$a^2 + b^2 = c^2 + 2ab×cos(t)$

Apply this to two vectors u and v in n-space. The length of u is its distance from the origin, which is found by the Pythagorean Theorem once again. If u is [3,1,7] then its length is sqrt($3^2+1^2+7^2$) = sqrt(9+1+49) = sqrt(59). But we're interested in a^2, the length squared, which is 59. In general, this is written as the dot product u.u, which is the coordinates of u times the coordinates of u, all added together.

In the same way, the length of v is sqrt(v.v). Of course we want b^2, which is v.v.

The segment from u to v is u-v, so square those coordinates and add them together to get c^2. Thus c^2 = (u-v).(u-v), which is the sum over $(u_i-v_i)^2$. Expand this to get u.u + v.v + 2×u.v. Subtract a^2+b^2 and you are left with 2×u.v. Apply the law of cosines, and u.v = cos(t) times the length of u times the length of v. It's a beautiful result.

As a corollary, u and v are perpendicular iff u.v = 0. This inspires the next definition.

A matrix M is orthogonal if every pair of rows is perpendicular. The dot product of any two rows is 0.

In 2 dimensions the two lines are at right angles. Two vectors typically define a parallelogram, but in this case they define a rectangle.

In 3 dimensions, the first 2 vectors determine a rectangle in space, and the third vector is perpendicular to the first two. It makes a perfect box, though it may be tilted relative to your coordinate system. The volume of such a box is length times width times height. If the rows of the matrix are u v and w, that is, the three vectors are u v and w, then the volume of the box is the square root of u.u times v.v times w.w. Here is a sample 3 by 3 orthogonal matrix. The volume of the box spanned by these vectors is sqrt(4×11×66) = sqrt(2904).

2	-1	1
1	3	1
4	1	-7

Determinant equals Volume

If a matrix M is orthogonal it defines a box in space. The volume of this box, squared, is the sum of $M_i.M_i$, over all the rows of M. And yet, this is the determinant of M times M transpose. Evaluate $M \times M^T$ and note that it is a diagonal matrix, with $M_i.M_i$ on the diagonal. It is easy to see that this matrix product is a diagonal matrix because the columns of M transposed are the rows of M, and the (i,j) entry in $M \times M^T$ is $M_i.M_j$, which is 0 when i is not equal to j, and the length of M_i squared when i = j. Thus $\det(M \times M^T)$ is volume squared. Since $\det(M^T) = \det(M)$, $\det(M)$ squared is volume squared. Therefore, magnitude of determinant equals volume, as long as M is orthogonal. With this in hand, let's step through 1 2 and 3 dimensions, and you can extrapolate from there.

In one dimension, the absolute value of the determinant is the volume. The determinant is the single entry of M, which is the distance to the origin, which is the length of the line segment, which is "volume" in one dimension.

Move on to two dimensions, where "volume" is area. Remember our example of u and v in the plane, where u is the vector from the origin to [2,1] and v is the vector from the origin to [1,2]. Here u is the floor of the parallelogram, while v slants up to touch the roof. (In this case the shape is a rhombus, since u and v have the same length.) If a parallelogram is a pushed over rectangle, then we are going to push it back. Slide the roof down and to the left, but make sure it is always moving parallel to u, parallel to the floor. The area of the parallelogram is base times height, and the height is not changing, so the area doesn't change either. Stop when the shape becomes a rectangle. The vector v has now rotated, so that it forms a right angle with u. It is shorter than it was, but that's ok. Meantime, what has happened to our determinant? As the endpoint of v slides back, parallel to u, the difference is some multiple of u. Therefore some multiple of u has been subtracted from v. This does not change the determinant. The shape has been turned back into a rectangle, and the determinant is unchanged. At this point the determinant equals the area, and we're done.

The area of the parallelogram equals the area of the rectangle that contains it. This is because triangle EJH is congruent to triangle GMI, using leg hypotenuse, or angle side angle, or side angle side, as you prefer.

Now move on to 3 dimensions. The shape spanned by 3 vectors is a parallelotope. This is a pushed over box, so push it back. Slide the ceiling back until it is directly over the floor, maintaining a constant height between the floor and ceiling at all times. The walls are now at right angles to the floor and the ceiling. This does not change height, or volume, and since multiples of the first two vectors are subtracted from the third, the determinant does not change either. Now it looks more like a box, but it could still be a parallelogram in cross section. Push one wall back until it is directly across from the other wall. This does not change volume, or determinant. Finally, the box is rectangular, while the determinant remains unchanged. Now that the vectors are orthogonal, determinant and volume agree. Since "pushing back" did not change the volume or the determinant, the determinant of M gives the volume of the parallelotope spanned by M.

It is hard to visualize this in higher dimensions, but the same proof works. Slide the "top" in n-1 dimensions until it rests directly over the "bottom", then, proceed through the other dimensions to obtain a rectangular box. Therefore, the absolute value of the determinant of an n by n matrix gives the volume of the parallelotope spanned by the rows of that matrix.

The Gram Schmidt Process

If you are looking for an algorithm to "push" the parallelogram back into a rectangle, as described in the previous section, then the Gram Schmidt process does the trick.

Start with a set of n independent vectors that span n-space. By induction, assume the first n-1 independent vectors have been transformed into orthogonal vectors spanning the same subspace. Call these orthogonal vectors x_1 x_2 … x_{n-1}. Let y be the next independent vector. For each i in 1 to n-1, subtract $x_i \times x_i$. y over $x_i.x_i$ from y. Call this new vector z, or if you prefer, x_n.

$$z = y - ((x_1.y)/(x_1.x_1)) \times x_1 - ((x_2.y)/(x_2.x_2)) \times x_2 - ((x_3.y)/(x_3.x_3)) \times x_3 - …$$

Verify that $z.x_i$ is 0 for each i in 1 to n-1. (If you're doing this over the complex numbers, where the dot product does not commute, then show $x_i.z = 0$ instead.)

$$z.x_1 = y.x_1 - ((x_1.y)/(x_1.x_1)) \times (x_1.x_1) = y.x_1 - x_1.y = 0$$

$$z.x_2 = y.x_2 - ((x_2.y)/(x_2.x_2)) \times (x_2.x_2) = y.x_2 - x_2.y = 0$$

$$z.x_3 = y.x_3 - ((x_3.y)/(x_3.x_3)) \times (x_3.x_3) = y.x_3 - x_3.y = 0$$

…

Since z and y differ by a linear combination of x vectors, the same space is spanned. Thus, z becomes the next vector in the orthogonal set.

Notice that z is a shift of y, sliding the roof into position over the floor. The determinant does not change.

Here is a simple example in two dimensions. Let v be the vector [4,0], pointing along the x axis, and let w = [1,1], pointing up and to the right. Now v and w form the bottom and left side of a parallelogram. The Gram Schmidt process pushes this parallelogram back into a rectangle. Subtract w.v/v.v times v, or ¼v, from w, hence w becomes [0,1]. Now w points straight up, and the vectors are orthogonal.

v = [4,0]. w = [1,1].

w.v = 4

v.v = 16

w.v/v.v = ¼

w - ¼v = [1,1] - ¼[4,0] = [1,1] - [1,0] = [0,1]

Gram Schmidt can be applied to a countable infinite basis as well. Shift each vector, one after another after another, until the entire basis is transformed.

The Shoelace Formula

Let s be an ordered set of points in the plane that defines a simple closed polygon. A segment joins the first point in s to the second, then another segment joins the second to the third, and so on, like dot-to-dot. The shape is closed as a segment joins the last point to the first.

Write the x coordinates of the points in one column and the y coordinates in another. Then multiply each xi by y_{i+1}, and each y_i by x_{i+1}. The former terms are added and the latter are subtracted. Divide the result by 2 to find the area of the polygon. This is the shoelace formula.

If you write down the coordinates in columns, and connect the items that are to be multiplied, the picture looks like shoelaces. The laces that slope down and to the right represent positive products and the laces that slope down and to the left represent negative products. Unlike most shoes, two extra laces connect the bottom entries with the top, in a wrap-around fashion, thus joining the last point in s with the first. I'll illustrate using 4 points that form a square. Actually it looks like a baseball diamond, but that's still a square. Run the bases in order, like hitting a home run.

x	y
0	-1
x	
1	0
x	
0	1
x	
-1	0
x	

The x coordinate of the last point, x_4 is multiplied by y1. The y coordinate of the last point, y_4, is multiplied by x1.

Left to right products: $(0{\times}0 + 1{\times}1 + 0{\times}0 + {-}1{\times}{-}1)$

Right to left products: $+({-}1{\times}1 + 0{\times}0 + 1{\times}{-}1 + 0{\times}0)$

The positive products add up to 2 and the negative products add up to -2. Subtract these to get 4. Then divide by 2 to get the area of the polygon, which is 2. This agrees with the area of the square, computed by traditional means. Now we are ready for the proof.

Add the constant c to all the x coordinates. You add cy_2 to the total, courtesy of x_1y_2, but you also subtract cy_2 from the total courtesy of y_2x_3. This holds for each yi, hence the total is unchanged. Slide the polygon to any x and y coordinate, and its area does not change, nor does the shoelace formula.

We get the same result by adding a constant to each of the y coordinates.Start with a convex polygon. This is a shape with no inlets or coves. Shift the polygon so that the origin is inside. We can do this by choosing some point (w,v) inside the polygon. Then subtract w from each vertex x coordinate, and subtract v from each vertex y coordinate.

The point (w,v) then maps to (0,0). Draw two segments from the origin to two adjacent vertices in the polygon. These vectors form a triangle with the edge of the polygon. They also span a parallelogram whose area is twice the triangle. The area of the parallelogram is given by the determinant, which is one of the shoelace cross products. Half the sum of these cross products is the area of the triangles in the polygon, which gives the area of the polygon. The shoelace formula agrees with the area.

We need to watch our signs however; determinants can be negative. Assume the points run counter clockwise. For any triangle, apply a rotation about the origin, until the base lies on the positive x axis. This does not change the area of the triangle, nor the parallelogram that contains it. As you pivot about the origin, the motion is continuous in the angle t. Each vector is a continuous trig function of t, and the determinant is a continuous function of t. It is always the area of the parallelogram, or minus the area of the parallelogram, but it cannot suddenly jump from one to the other. The sign does not change. So evaluate the determinant when the base runs along the x axis, and the next vector points up into the upper half plane, having a positive y coordinate. The determinant is positive, giving the true area of the parallelogram. This holds for each triangle around, and the formula is correct.

Just for grins, run around the polygon clockwise instead of counterclockwise. This flips the shoe upside down. Instead of x3×y4 - y3×x4, you have x4×y3 - y4×x3. Each determinant is negated. Run the perimeter clockwise and the shoelace formula gives minus the area of the polygon. Look again, as we run the bases in reverse, from home to third to second to first and back home.

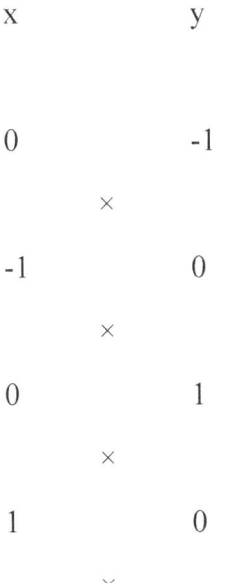

x	y
0	-1
×	
-1	0
×	
0	1
×	
1	0
×	

Left to right products: $0 \times 0 + -1 \times 1 + 0 \times 0 + 1 \times -1$

Right to left products: $-1 \times -1 + 0 \times 0 + 1 \times 1 + 0 \times 0$

$(-2 - 2) / 2 = -2$

Next assume s is not convex. Travel around the polygon and find a cove, where the polygon bends inward and then turns back out again. Let v and w be the points where the cove begins and ends. If s looks like a crescent moon, v and w could be the tips of the crescent. If s is a star, v and w could be adjacent points on the star. In any case, the segment vw closes up the cove.

Translate the polygon so that the origin is midway between v and w. The original polygon s does not travel directly from v to w, it passes through at least one other vertex to get there. A new polygon s_2 takes the direct route from v to w, and then continues the path of s from w back around to v. This bypasses the cove. A third polygon s_1 travels from v to w directly, and then reverses the original path of s from v to w. This is the perimeter of the cove. The path from w to v, in s1, is opposite to the path from v to w, in s. Note that the area of s is the area of s_2 minus the area of s_1.

Even with the new segment v to w added, s_1 and s_2 have fewer edges than s. By induction on the number of edges, the shoelace formula works for s_1 and s_2.

Delete the connection from v to w, in either s1 or s2. This cuts two of the shoelaces, and deletes the two products associated with v above w. But what do v and w look like? The vectors from 0 to v and from 0 to w are antiparallel. Something like a,b and -a,-b. Their determinant is 0. The shoelace formula is the same, whether this determinant is included or not. Start with w and walk around to v, and list the coordinates in columns. There is no need for the connection from v to w, from the bottom of the shoe back around to the top, so I'm going to leave it out.

Write the shoelace formula for s_2, starting with w and walking around counterclockwise to v, without the connection from v to w. Below this, write the shoelace formula for minus the area of s_1. Start with v and walk around clockwise to w, leaving out the connection from w to v. Now one shoe sits above the other. The first shoe starts with w, and ends with v. The second shoe starts with v and ends with w. Paste the two shoes together at the vertex v, and let w, at the top of the first shoe, be the same point as w at the bottom of the second shoe. This is now the shoelace formula for s, starting at w and walking all the way around and back to w. The shoelace formula gives the area of s_2 minus the area of s_1, which agrees with the area of s. That completes the proof.

A special case is the area of a triangle. Assume the vertices of a triangle are located at (x1,y1), (x2,y2), and (x3,y3). The area of the triangle is as follows:

$| (x1y2 - y1x2) + (x2y3 - y2x3) + (x3y1 - y1x3) | / 2$

Let the vertices of the triangle be (0,0),(2,0), and (2,3).

Area is $((0 \times 0 - 2 \times 0) + (2 \times 3 - 2 \times 0) + (2 \times 0 - 3 \times 0))/2 = 3$.

Orthonormal and Rotations

An orthogonal matrix is orthonormal, (also called unitary), if the length of each vector is 1. $M_i . M_i = 1$, and $M_i . M_j = 0$ for j not equal to i. The box is a perfect cube, and its volume is 1.

The following matrix in 2 dimensions is orthonormal. The dot product of each row with itself is 1, and the dot product of the first row with the second is 0. The two vectors span a perfect square (albeit tilted) in the plane. Plot these vectors in the xy plane and see for yourself.

3/5   4/5 $(3/5) \times (3/5) + (4/5) \times (4/5) = 1$, and

-4/5   3/5 $(3/5) \times (-4/5) + (4/5) \times (3/5) = 0$.

The transpose of a matrix reflects through the main diagonal, and turns rows into columns. Multiply M by M^T and get the identity matrix. Thus the determinant of M is ±1.

Let M be an orthonormal matrix and view it as a linear transformation. In other words, the first coordinate is mapped to the top row of M, the second coordinate is mapped to the second row of M, and so on. Since the map is linear it is completely determined by the vectors in M. To illustrate, let x be a point in n-space with coordinates x1 x2 x3 … xn. Multiply x1 by the first row of M, x2 by the second row of M, and so on; then add up these modified rows to get the image of x. The matrix product x×M becomes f(x).

Continue our earlier example, and multiply [x1,x2] by our orthonormal matrix in the plane. This is a function that takes an input vector on the left and produces an output vector on the right.

$[x_1, x_2]$

×

3/5   4/5

-4/5   3/5

=

(3/5)x1 - (4/5)x2

(4/5)x1 + (3/5)x2

Now consider two vectors x and y. They are mapped to x×M and y×M respectively. Compute the dot product of their images. It will magically come out the same as x.y. Let's see why.

The first coordinate of the image of x is the dot product of x with the first column of M. The first coordinate of the image of y is the dot product of y with the first column of M. Multiply these two expressions together and expand. For i not equal to j, we find terms like $x_i y_j M_{i,1} M_{j,1}$. This looks like a mess, especially when you repeat it for all the other columns of M, but with proper regrouping it all clears up. The common factor $x_i y_j$ is multiplied by the sum of $M_{i,k}M_{j,k}$ as k runs from 1 to n. This is the dot product of rows i and j in the matrix M. Since M is orthogonal that drops to 0. We can forget about the cases where i is not equal to j.

When i = j, x_iy_i is multiplied by the sum of $(M_{i,k})^2$ as k runs from 1 to n. This is the ith row dotted with itself, and in an orthonormal matrix, that is 1. The dot product of the image of x with the image of y is the sum of x_iy_i, which happens to be the dot product of x and y. Therefore, M preserves dot product.

Look at an example in 2 dimensions. The input vectors are [x1,x2] and [y1,y2], and their dot product is x1y1 + x2y2. The orthonormal matrix is as above, but for notational convenience I will set r = 3/5 and s = 4/5. Run these two vectors through the orthonormal matrix, then take the dot product.

[x1,x2]

$\times$

r   s

-s   r

=

rx_1 - sx_2

sx_1 + rx_2

y_1 y_2

$\times$

r   s

-s   r

=

ry_1 - sy_2

sy_1 + ry_2

$[rx_1 - sx_2, sx_1 + rx_2]$ dot $[ry_1 - sy_2, sy_1 + ry_2] =$

$(rx_1 - sx_2) \times (ry_1 - sy_2) + (sx_1 + rx_2) \times (sy_1 + ry_2) =$

$r_2 x_1 y_1 - sr x_2 y_1 - rs x_1 y_2 + s_2 x_2 y_2 +$

$s_2 x_1 y_1 + rs x_2 y_1 + sr x_1 y_2 + r_2 x_2 y_2 =$

$(r2 + s2) x_1 y_1 + 0 \times x_2 y_1 + 0 \times x_1 y_2 + (s_2 + r_2) x_2 y_2 =$

$1 \times x_1 y_1 + 1 \times x_2 y_2 = x_1 y_1 + x_2 y_2 = [x_1, x_2]$ dot $[y_1, y_2]$

Since length is determined by dot product, an orthonormal transformation preserves lengths. Line segments don't grow or shrink as you apply the transformation. Furthermore, angle is a function of dot product and lengths, as we saw earlier with the law of cosines, so angles are preserved as well. A shape is not bent or twisted as it is remapped; it is simply moved to a new location. The origin maps to itself, so M pivots space about the origin. This is called a rigid rotation in space. (Within this context, a rotation could also reflect space through a mirror.)

Conversely, let a matrix M implement a linear transformation that preserves lengths and angles. In other words, it preserves the dot product. Pre-multiply M by the first coordinate [1,0,0,...] to extract the first row. This must have length one, just like the unit vector that produced it, hence the first row of M has length 1. The same holds for the remaining rows of M.

The dot product of any two coordinate vectors is 0, example [1,0,0,0,...] with [0,1,0,0,...], hence the dot product of any two distinct rows in M is 0. This makes M orthonormal. Therefore, a matrix is orthonormal iff it preserves lengths and angles.

Since the composition of two such functions preserves lengths and angles, the product of two orthonormal matrices is another orthonormal matrix. In other words, orthonormal matrices are closed under multiplication, just as rigid rotations are closed under composition.

We already showed the inverse of an orthonormal matrix is its transpose, and the inverse of a rigid rotation preserves lengths and angles, hence the inverse, or transpose, of an orthonormal matrix is orthonormal. More formally, let x' and y' have dot product c, and let them come from x and y under M. Since M preserves dot product, x.y = c. Apply M inverse to x' and y' and get x and y, having the same dot product. Thus M inverse, or M transpose, preserves dot product, and is orthonormal. It follows that the rows are orthonormal iff the columns are orthonormal.

If a matrix is orthogonal, but not orthonormal, its transpose need not be orthogonal, as shown by this example. Also, its inverse, or its square, is not necessarily orthogonal.

M

4	2
-1	2

M^T

4	-1
2	2

M^2

14	12
-6	2

M^{-1}

0.2	-0.2
0.1	0.4

The inverse of a matrix with determinant 1 is its adjoint. If M is orthonormal its inverse is also its transpose. If M is orthonormal with determinant 1, the transpose of M is the adjoint of M. Each entry of M is its own cofactor. For example, delete the first row and column and take the subdeterminant, and you get the upper left entry back again.

If an orthonormal matrix has determinant -1 then the adjoint is minus the transpose. Delete the first row and column and take the subdeterminant, and you get minus the upper left entry back again.

There is a hierarchy of matrices as follows. Start with the ring of all matrices. Then the invertible matrices, with nonzero determinants, correspond to linear functions that can be reversed. Within this group are the matrices with determinant ± 1. These may turn a square into a long thin parallelogram, but volume is preserved. Within this group are the orthonormal matrices, the rigid rotations and reflections in space. Within this group are the orthonormal matrices with determinant 1. These are the rotations in space without the reflections. More on this below.

All matrices

Invertible

Determinant ± 1

Orthonormal

Rotations

Here is an orthonormal matrix that implements a reflection, rather than a rotation. The x and y axes switch positions, while the z axis stays put. This is a reflection through the plane x = y. Lengths and angles are still preserved. As with all reflections, the determinant is -1.

0	1	0
1	0	0
0	0	1

According to det2, the determinant is a polynomial in the entries of M. Thus, determinant is a continuous function of M. Yet in an orthonormal world, or even an invertible world, positive and negative determinants are separated by 0, which cannot be achieved. You can continuously spin an object about, this way and that, thus defining new rotations and new matrices, but each one will have determinant 1. You can't gradually slide from 1 to -1, from a rotation to a reflection. In the same way, if an object is reflected, you can rotate it around any axis, again and again, but it will still be reflected. The determinant cannot slide back from -1 to 1.

As intuition would suggest, there is a continuous path of rotations from the start position to any given rotation - a continuous path of matrices from the identity matrix to any orthonormal matrix with determinant 1. You realize this path when you turn the object with your hand to its new position. The coordinate system gradually moves, intact, towards its destination, until it finally coincides with M. That's what we mean by rotation, isn't it? The Earth doesn't suddenly jump from day to night; it continuously spins around. Let's try to build such a path of matrices.

In one dimension the only rigid rotation is [1]; the unit vector just sits there. The single reflection maps 1 to -1.

In two dimensions the rigid rotations spin the plane about the origin. The vector [1,0] is mapped to a point on the unit circle, and [0,1] is mapped to a point 90° ahead. If [s,t] is a point on the unit circle, the matrix has [s,t] in the top row and [-t,s] in the bottom row. All these linear maps are determined by some angle, t, and functions are composed by adding angles. Thus the rigid rotations in two dimensions commute. This is not the case in higher dimensions. Take a die, spin it about the x axis, then the y axis, and see how it lands. Start again, spin it about the y axis, then the x axis. The orientations are different.

Returning to 2 dimensions, let the top row of M be [s,t] as before, but let the bottom row be [t,-s]. This is still orthonormal, but the determinant is -1. It is a reflection. The y axis lands 90° behind the x axis, instead of 90° ahead. Still, any reflection can be continuously transformed into another reflection by rotating the basis through an appropriate angle. Slide s and t around the unit circle, cos(t) and sin(t) to be precise, and the matrix slides along, and the rotation or reflection slides along.

Interestingly, each reflection really is a simple mirror image, if you tilt the mirror properly. Start with the standard basis and map y to -y, a reflection through the x axis. This is what you get when s = 1 and t = 0. Take this reflection and rotate it counterclockwise through an angle of t. You can accomplish the same thing by reflecting the plane through a line that meets the x axis at an angle of t/2. This is our tilted mirror. Choose any point on the unit circle, representing an angle of w. The first path takes w to -w, then adds t. The second path adjusts w according to the mirror (subtracting t/2), reflects through the mirror, then tilts back relative to the x axis. This gives (w-t/2)×(-1)+t/2. This is algebraically equivalent to -w+t.

Because rotations and reflections are represented by matrices, there is a connecting path of matrices for rotations, or reflections, in 2 dimensions. Adjust t continuously from 0 to your destination angle. For higher dimensions, proceed by induction on n.

The top row of M, call it v, must map continuously onto w, the top row of the identity matrix. Let P be the plane containing v and w, and spin v around to coincide with w, keeping the rest of space perpendicular to v and w at all times. This is a continuous change in t, in 2 dimensions, as described above. The rest of M is mangled in some way, but the top row has become w, and the remaining rows are perpendicular to w, just as they were perpendicular to v at the start. Rotations aren't going to change that. The matrix remains orthonormal at all times.

Except for a 1 in the upper left, the first row and column of M are all zeros. The rest of M is orthonormal, and spans the subspace perpendicular to w. Furthermore, its determinant is still 1. By induction one can map the rest of M onto the identity matrix. Therefore, every rotation can be continuously transformed into the identity map, and every pair of rotations is path connected.

If the matrix is a reflection, apply the same procedure. You can always morph to the modified identity matrix with -1 in the lower right. At the base of this inductive argument, any reflection in 2 dimensions can be continuously transformed into [1,0] [0,-1]. Thus all reflections are path connected.

In higher dimensions, reflections are no longer simple mirror images. You can't just tilt the mirror; a pre or post rotation is necessary. To illustrate, reflect 3 space through the xy plane, then spin it counterclockwise 90° around the z axis. These operations commute, so do them in either order. If this were a simple reflection the positive z vector would be reflected through a mirror, and would point down the negative z axis. The mirror has to be the xy plane. At the same time the positive x axis has become the positive y axis, hence the mirror is the plane x=y. It can't be done with mirrors, as was the case with 2 dimensions.

Cross Product

Most textbooks define the cross product of two vectors in 3 space and leave it at that. However, there is a more general definition. The cross product is a function that takes an ordered set of n-1 vectors in n space and produces a vector that is perpendicular to all of them. When two vectors are crossed in 3 space, the symbol × is sometimes used as a binary operator, as in a×b. Don't confuse this with the dot product a.b.

To compute cross product, build a square matrix M as follows. Fill in the first n-1 rows of M with the n-1 vectors in the set, in order, and leave the bottom row blank. Now we're ready to compute the last row of M. This vector will be the cross product.

Let $M_{n,i}$ be the ith cofactor. In other words, ignore the last row and ith column, take the determinant of what remains, and multiply by -1 if (n+i) is odd. Do this procedure for each column. The last row of M becomes the cross product. Here is a 3 dimensional example.

3	3	3
7	0	5
?	?	?

The bottom row is (3×5-3×0), -(3×5-3×7), (3×0-3×7).

The full matrix is

3	3	3
7	0	5
15	6	-21

We computed the three cofactors and found a cross product of (15,6,-21). This vector, call it v, is perpendicular to the other two vectors. The dot product of v with either the first or the second row is zero. Thus, the cross product is a handy way to find a vector perpendicular to the plane determined by two other vectors. Let's prove this in general.

Let v be the cross product of (n-1) vectors in n space. Now consider the dot product of v and the jth row of M. Compare this to calculating the determinant of M with the jth row copied onto the last row. Recall that the ith component of the cross product is its cofactor. Thus taking the dot product of the cross product with the jth row is equivalent to taking the determinant of M, with the jth row copied onto the last row. This gives a duplicate row, hence that determinant is 0. Therefore, v is perpendicular to all the vectors in the cross product, and the subspace spanned by those vectors.

Let v be the last row of M. This makes M into a basis. Apply a rigid rotation that moves v to the z axis, where z is the last dimension in our vector space. The remaining vectors are still perpendicular to v. Thus M has all zeros in the last row and column, except for $M_{n,n}$. Let $M_{n,n}$ = d. Now d is the subdeterminant of the rest of M. As shown earlier, the subdeterminant is the volume of the cell spanned by the n-1 vectors within their subspace. The length of the cross product equals the volume of the vectors that built it. In our example, the cross product was (15,6,-21), which has length sqrt(702). You could show that the parallelogram spanned by those two vectors has area sqrt(702) by traditional geometric methods, but it would be much more difficult.

If the vectors are permuted, the cross product might change sign; that's it. This does not change the line determined by the cross product, nor the magnitude of the vector, which indicates volume. Therefore most people don't worry about the order of the vectors. It is often sufficient to say, "Take the cross product of these vectors." But sometimes the sign is important, and when it is, you must remember that a×b = -b×a.

In 3 dimensions, you can move the row of question marks to the top. This is an even permutation of rows, and does not affect the sign. Thus, most text books define cross product with the question marks at the top.

?	?	?
3	3	3
7	0	5

The triple scalar product of three vectors in 3 space is a.b×c. In other words, take the cross product of two of them and dot with the third. Since dot product commutes, a.b×c = b×c.a.

Arrange these three vectors in a 3×3 matrix and note that the triple scalar product is the determinant. As an immediate corollary, a.b×c gives the volume of the cell spanned by those three vectors. Also, we can rearrange the three vectors, which simply permutes the rows in the matrix. Thus a.b×c = b.c×a = c.a×b. The remaining permutations, the odd permutations, multiply the result by -1.

Vandermonde Matrix

Throughout this section, rows and columns will be numbered starting with 0. It makes the math easier.

The matrix M is a Vandermonde matrix if $M_{i,j} = (M_{i,1})^j$. Here is a simple example.

1	1	1	1
1	2	4	8
1	3	9	27
1	5	25	125

As you can see, the jth column is the second column (having index 1) raised to the j power. Let x be this second column, as a column vector, again starting with x^0 at the top, so that $M_{i,j} = (x_i)^j$. In fact, let the elements of x be unknowns, variables that you can fill in later. Thus the determinant of M becomes a polynomial in n variables, x_0 through x_{n-1}. Example off 4 by 4 is

1	x_0^1	x_0^2	x_0^3
1	x_1^1	x_1^2	x_1^3
1	x_2^1	x_2^2	x_2^3
1	x_3^1	x_3^2	x_3^3

Evaluate the Vandermonde determinant via det2(M), the sum of permutation products. Consider the expression x_m-x_k for any m > k. For notational convenience, a = x_k and b = x_m. Consider 0 = i < j < n. For every permutation product with $a_i \times b_j$ there is another with $a_j \times b_i$. These products come from permutations with opposite parity, so subtract them, pull out the common factors from the other rows, and pull out $a_i b_i$. That leaves $a_{j-i}-b_{j-i}$, which is divisible by b-a. The entire determinant, as a polynomial, is divisible by x_m-x_k. This holds for every m > k.

Multivariable polynomials over the integers exhibit unique factorization. (This will be proved later.) Thus the unique factorization of the determinant includes x_m-x_k for each m > k. Furthermore, their product gives a polynomial of degree $(n^2-n)/2$. Now consider any permutation product in the determinant of M. It's degree is 0+1+2+...+n-1 = $(n^2-n)/2$. The linear factors identified thus far create a polynomial that divides (goes into) the determinant, and has the same degree, hence it can only be off by a constant.

Run down the main diagonal and note that the product of xii appears in the determinant with coefficient 1. Multiply the binomials together: (x1-x0) × (x2-x0) × (x2-x1) × (x3-x0) × ..., and one of the resulting terms, taking the first variable from each binomial, gives the same expression, with a coefficient of 1. Therefore the determinant of M is the product of x_m-x_k, for 0 = k < m < n.

In our example, the determinant is (2-1) × (3-1) × (5-1) × (3-2) × (5-2) × (5-3) = 48.

As a corollary, any set of distinct Vandermonde vectors, drawn from an integral domain, is linearly independent. The determinant is the product of nonzero differences, and is nonzero.

A Matrix of Matrices

As described earlier, a matrix is an object that can have other objects inside it. Consider a matrix with other matrices inside it. For example, let the class of matrices be 3 by 3, and let the internal elements be matrices that are 2 by 2. Here is an illustration.

8	0		3	4		8	2
5	1		7	1		6	7
3	4		7	8		6	2
5	5		3	0		1	1
5	4		2	6		1	6
9	2		0	7		4	3

These can be viewed as compound matrices, as described above or as simple 6 by 6 matrices. There is virtually no difference.

Consider the sum of matrices A+B. Add the submatrices together by adding their numeric elements in the usual way, or just add up the numbers straight across; the result is the same.

Multiplication is a bit trickier. Take the dot product of the first row of submatrices of A, with the first column of submatrices of B. Use matrix multiplication to evaluate A1,i times Bi,1, then matrix addition to add up the three products. The upper left entry in this submatrix is 2 numbers dotted with 2 numbers, plus 2 numbers dotted with 2 numbers, plus 2 numbers dotted with 2 numbers. This happens to be the same as the top row of A dotted with the left column of B, viewed as 6 by 6 matrices. This carries all the way across; hence, multiplication is the same whether the matrices are simple or compound.

Be careful though; the determinant of the determinant is not the determinant. The following 4 by 4 matrix has a determinant of x-y. However, when viewed as 4 2×2 matrices, the product of the upper left block times the lower right block leaves x and y on the cutting room floor.

0	1	0	0
0	1	1	1
0	1	x	y
1	0	0	0

That said, there are branches of mathematics where the determinant of the determinant is the determinant. In algebraic number theory, the determinant implements the norm, and yes indeed, the norm of the norm is the norm. The norm of an extension will be described later.

Of course, entire books have been written about matrices. This is merely a foundation - sufficient to describe matrices as a ring or a group. We will no doubt return to this topic in more detail later on. For now, let's develop some additional structures, classes if you will, that will serve as examples of groups and rings.

Complex Numbers and the Gaussian Integers

Within the real numbers, a negative number has no square root. In particular, x^2+1 has no solution. So we simply imagine one. In fact the solution is called i, for an imaginary number. (Electrical engineers call it j, because i stands for current. Why don't they use c for current? It is because c stands for capacitance. In addition, l stands for inductance. It's a whole nother world out there.)

With i in place, a complex number has two components, one real and one imaginary. It looks like a+bi. Addition is performed per component, and multiplication distributes across terms. Thus (a+bi) × (c+di) = (ac-bd) + (ad+bc)i.

Complex numbers are often plotted in the complex plane, where real numbers run along the x axis and imaginary numbers (multiples of i) run up the y axis. Thus 3+4i is in the first quadrant, and is 5 units distant from the origin by the Pythagorean theorem.

The points with integer coordinates in the complex plane are called the Gaussian integers, because Gauss used them to prove a number of theorems that were formerly inaccessible. You will see the name Gauss over and over again; he was indeed The Prince of Mathematics. I've already referred to Gaussian elimination, which solves long-standing problems in linear algebra. You may have seen Gauss's theorem in multi-variable calculus, or heard of gauss as a measure of magnetism. This is just the tip of the iceberg. One of my professors remarked, "Reading his Disquisitiones Arithmeticae, that's reason enough to learn German."

Things are different in the Gaussian integers. You can see this right away, because 17 is no longer prime. In fact, 17 = (4+i) times (4-i). We'll explore prime factorization later, but first we need a few more homomorphisms.

Four has two square roots, 2 and -2. In the same way, -1 has two square roots, i and -i. When a root is "imagined" into existence, it is essentially indistinguishable from any other root of the same polynomial. Replace i with -i throughout and nothing changes. This is called conjugation. It reflects the complex plane through the x axis, like a mirror. It is worth verifying this automorphism. Consider the sum, or product, of two complex numbers. The conjugates of a+bi and c+di are a-bi and c-di. Add these together and you do indeed get the conjugate of a+b + (c+d)i. Perform a similar observation for multiplication. Finally, every a+bi has a unique preimage, namely a-bi, so the homomorphism is actually an automorphism.

Another useful homomorphism is the square of the distance to the origin. This is called the norm, and is sometimes written with vertical bars, as in |a+bi|. By the Pythagorean theorem, $|a+bi| = a^2+b^2$. If you use conjugation, it is easy to show that norm is a homomorphism that respects multiplication. First show that |a+bi| = a+bi times a-bi. Then, given two complex numbers s and t, step through the following equivalences.

|st| = st × conjugate(st)

= s×t×conjugate(s)×conjugate(t) (conjugation is a homomorphism)

= s×conjugate(s)×t×conjugate(t) (multiplication is commutative and associative)

= |s| × |t|

The norm of st is the norm of s times the norm of t, and norm is a homomorphism that respects multiplication.

The norm makes it easy to compute the quotient s/t. Let r = s×conjugate(t)/|t|. Multiply r by t and get s×|t|/|t|, or s. Thus the quotient is s times t conjugate divided by the norm of t. Since the norm of t is 0 only if t is 0, this is well defined.

A Gaussian integer s is invertible, with st = 1, only if |s|×|t| = 1, hence |s| is also invertible. Norms are positive integers, so |s| = 1. Conversely if |s| = 1 then s times its conjugate is 1, and s is invertible.

Look at the grid in the complex plane; the only points that are a distance of 1 from the origin are 1, -1, i, and -i. These are the invertible units in the Gaussian integers.

Let p = 2+i, hence |p| = 5. If p is not prime, then write p = st. Take norms, and 5 = |p| = |s|×|t|. Yet 5 is prime in the integers. One of the two norms is 5 and the other is 1. Say |s| = 5 and |t| = 1. The norm of t is 1, t is a unit, (1, -1, i, or -i), s is an associate of p, and p = 2+i, is prime after all. You can tell, using the norm homomorphism, that certain Gaussian integers, such as 3+2i (with norm 13), are prime. We will explore this topic in more detail once unique factorization has been established.

This is not the first time a structure has been extended by adjoining a new element, to satisfy a polynomial that had no solution before. The Greeks understood the integers, and the fractions (ratios) thereof, which we now call the rational numbers. They wondered whether some fraction, perhaps the quotient of two 50 digit numbers, might be precisely the square root of 2. In other words, a rational solution to x^2-2. Euclid made a compelling argument, if not a formal proof, that such a fraction could not exist. Yet Pythagoras showed that such a distance could be constructed using a right triangle, so what to do? Simply adjoin the square root of 2 to the integers. This was the first "irrational" number, i.e. not a ratio.

Let q be a solution to x^2-2. As with i, -q is also a solution, and conjugation maps q to -q. If a and b are rational numbers, then conjugate(a+bq) = a-bq. Verify that this is an automorphism, just as you did for complex numbers. Then let the norm |a+bq| = a^2-$2b^2$ = (a+bq) times (a-bq). The norm of s is s times its conjugate. Verify that norm is a multiplicative homomorphism, using the steps outlined above.

Unlike the complex numbers, this extension has many units. There are many numbers wherein a^2-$2b^2$ = 1. For example, 17+12q times 17-12q = 289 – 2 × 144 = 1.

These two examples share a common theme. There is a structure, technically a ring, and a polynomial p(x) that has no solution. Simply adjoin a root of p(x), as though it had been there all along. Conjugation is an automorphism that maps one root to another, and the norm of s is the product of the conjugates of s, including s.. This will become clearer once we are able to define it formally. In the meantime, just think of the reals stepping up to the complex numbers by adjoining i. This is a system that you already know, and it captures the essence of a ring extension.

Another nice property of the complex numbers is closure. If you join the square root of 2 to the rationals, you still don't have the square root of 3. There are still other polynomials without solutions. However, if you start with the reals, and bring in i, every polynomial has a solution. There is nothing else to do. This is the fundamental theorem of algebra, and there are many different proofs. I will present one based on galois theory, after we have a lot more machinery in place.

For notational convenience, the integers are denoted **Z**, the rationals are denoted **Q**, the reals are denoted **R**, and the complex numbers are denoted **C**. Thus **C** is **R** adjoin i, or **R**[i].

The integers mod p are denoted **Z**/p, as though the integers were divided by p, leaving 0 through p-1.

Eisenstein Integers

Let w be a root of x^2+x+1 = 0, and adjoin w to the integers. The quadratic formula shows w = ½(-1±sqrt(3)i). These two points are 1 distance away from the origin, and they have an angle of 120 degrees from the positive x axis. Bring in the point 1,0 and you have three points on the unit circle, spaced evenly at 120 degrees apart. We confirm this algebraically. We multiply (x^2+x+1) by (x-1) and get (x^3-1). Each of these three points satisfies x^3-1=0. Each of these three points is a cube root of 1. Of course 1 is a trivial cube root of 1. Hence this extension brings in the other two cube roots of 1. To avoid ambiguity, let w = ½(-1+sqrt(3)i), so that the conjugate w^2 = ½(-1-sqrt(3)i). Thus w lies above the x axis and w conjugate lies below.

Recall the Gaussian integers, where multiples of 1 and i tile the plane in a checkerboard pattern. In this extension, multiples of 1 and w tile the plane with parallelograms. Actually, each parallelogram is a rhombus, with sides of length 1. Cut the rhombus in half with a diagonal and find two equilateral triangles. This is illustrated by drawing a line from the origin to w+1 = ½(1+sqrt(3)i). Cut all the parallelograms in half in this manner, and tile the plane with equilateral triangles. Thus, these points are sometimes called triangular integers, or Eisenstein integers, after the mathematician who explored them.

Replace i with -i; this is complex conjugation and it is an automorphism on the entire (continuous) plane, as described earlier. Substitute in the formula for w, and conjugate(w) = -1-w. This reflects w through the x axis, and carries one cube root of 1 to the other. In general, conjugate(a+bw) becomes (a-b-bw).

Let the norm of s be s × conjugate(s) as usual. We already showed that norm, based on complex conjugation, is the square of the distance to the origin, and is a homomorphism that respects multiplication. The norm of s × t is the norm of s times the norm of t, across the entire complex plane. Apply this to a point that is expressed as a+bw. Multiply this by its conjugate and get a^2-ab+b^2. At first this looks like it might come out negative, but it is still the distance squared to the origin, and has to be a positive integer, or 0 if a = b = 0.

Once again the norm allows for division. The inverse of s is conjugate(s)/|s|.

The units are the points with norm 1, the points that lie on the unit circle. There are 6 of them, spaced 60 degrees apart. These are the three cube roots of 1 that we already know, and their opposites. Together they form the 6 sixth roots of 1.

If the norm of p is prime then p is prime. An example is 3+2w, which has a norm of 7. Use the proof shown above, writing 3+2w = st, whence 7 = |s|×|t|, |t| = 1, t is a unit, and 3+2w is prime after all. This means 7 is not prime; it is the product of 3+2w and 1-2w.

$(3+2w) \times (1-2w) =$

$3 + 2w - 6w - 4w^2 =$

$3 - 4w - 4w^2 =$

$3 - 4(w + w^2) =$

$3 - 4 ((1+w+w2) - 1) =$

$3 - 4 (0 - 1) =$

$3 + 4 =$

7

Quaternions

Start with the reals or the rationals if you like, or even the integers, and adjoin i, the square root of -1, as we did earlier. Then adjoin j and k, which are also square roots of -1. These three symbols do not commute; they multiply as follows.

i×j = k

j×k = i

k×i = j

j×i = -k

k×j = -i

i×k = -j

A typical quaternion looks like a+bi+cj+dk. Addition takes place per component, and multiplication distributes across components. The product of a+bi+cj+dk times e+fi+gj+hk looks like this.

(ae-bf-cg-dh) + (af+be+ch-dg)i + (ag+ce+df-bh)j + (ah+de+bg-cf)k

Take a moment to show that multiplication is associative. You only need demonstrate this per component; the distributive property does the rest. **R** is associative, so we only need look at the adjoined elements. Consider i j or k, times i j or k, times i j or k, 27 possibilities. For example, (ij)i = ki = j, and i(ji) = -ik = --j = j. Check this for all 27 combinations, and the quaternions are associative. Of course, they are not commutative, since ij is not the same as ji.

The conjugate of a+bi+cj+dk, written conjugate(a+bi+cj+dk), is a-bi-cj-dk. Unfortunately, this function does not respect multiplication. Start with ij = k and conjugate, giving (-i)(-j) = k; rather than -k. Thus conjugation cannot be used to verify the properties of norm, as it was earlier.

The norm of s, written |s|, is s times conjugate(s), or $a^2+b^2+c^2+d^2$, which is the distance squared from s to the origin in 4 space. Norm is a homomorphism that respects multiplication, but since conjugation does not respect multiplication, we have to use brute force algebra to prove it. Review the formula above for the product of two quaternions. Square each component, and add them up. The result is the same as (a2+b2+c2+d2) × (e2+f2+g2+h2).

Once again the norm makes it easy to divide, as was the case with complex numbers. The inverse of t is conjugate(t)/|t|. But with the quaternions, you can divide on the left or on the right, i.e. multiply s by t inverse on the left or the right. In general the answers will be different.

If s is invertible then st = 1, |s|×|t| = 1, and |s| = 1. Conversely if |s| = 1 then s times its conjugate is 1, and s is invertible. In summary, s is a unit iff |s| = 1.

Ask which quaternions satisfy a2 + b2 + c2 + d2 = 1. These are points on the unit sphere in 4 space. There are 8 units: ±1, ±i, ±j, and ±k.

If p is a quaternion, and |p| is prime, then p is prime (stretching the definition of prime just a bit). Apply the norm homomorphism, just as we did for the gaussian integers. Illustrate with p = 2+i+j+k, and suppose p = st. Note that the norm of p, that is, p times its conjugate, is 2+i+j+k times 2-i-j-k, or 4+1+1+1, or 7. Take norms, and 7 = |s|×|t|. Yet 7 is prime in the integers. One of the two norms is 7 and the other is 1. Say |s| = 7 and |t| = 1. The norm of t is 1, t is a unit, s is an associate of p, and p is prime after all. Either s = p, or some variation of p, such as -2-i-j-k. These variations are called associates, like 5 and -5 in the integers. In the Gaussian integers, every number other than 0 has 4 associates. In the quaternion integers, every number other than 0 has 8 right associates. Multiply s by ±1, ±i, ±j, or ±k to find the 8 associates. There are also 8 left associates, which are not always equal to the 8 right associates. Multiply 1+2i by units on the right, and the left, and compare the two sets.

Since 2+i+j+k turned out to be prime, the factorization of 7, in this ring, is the product of two primes 2+i+j+k times 2-i-j-k. These are two primes that "lie over" 7.

Although conjugation did not turn out to be an automorphism, there is a circular automorphism of period 3. Map i to j, j to k, and k to i. Start with ij = k, for example, and apply the map, and get jk = i, which is true. Verify the other combinations and you're home free.

If u is a quaternion a+bi+cj+dk, square u and get a real number + 2au. Thus u satisfies a quadratic equation over the base ring.

Half Integer Quaternions

The Gaussian integers define a checkerboard in the plane; picture the quaternions as a grid of cubes in 4 dimensions. The base cube is spanned by 1, i, j, and k, and has volume 1. Now include all the centers. For instance, the center of the base cube is ½+½i+½j+½k. Do this across space, so that the components are either all integers or all half integers.

Addition causes no trouble. The sum of two points has all integers or all half integers.

Multiplication is a little trickier. Integers times integers produce integers, so multiply a half integer quaternion by a "regular" quaternion. If a component in the latter is even, it contributes an integer to each component in the product. An odd entry, such as 7j, adds a half integer to each component in the product. At the end of the day all four components are integers or half integers.

Finally multiply two half integer quaternions together. Write each coefficient as some integer over 2. Reduce the numerators mod 4, so that each is 1 or -1. Negate each factor if necessary, so that the real component of each is 1 mod 4. If a factor starts out 1-i... mod 4, premultiply by i, so that the first two coefficients are 1 mod 4. The remaining coefficients, on j and k, may be 1 or -1 mod 4. That's 16 possibilities. Try them out and see. The result is always 0 mod 4 or 2 mod 4 across the board. Bring in the denominator, ½×½ = ¼, and find integers or half integers. The product is well defined.

Here is one of the 16 combinations.

½ (1+i+j+k) × ½ (1+i+j-k) =

¼ { (1-1-1+1) + (1+1-1-1)i + (1+1+1+1)j + (-1+1-1+1)k } =

0 + 0i + 1j + 0k = j

Consider the norm of one of these centers, an element with half integer coefficients. The square of an odd number is 1 mod 4. Add four of these squares together to find a multiple of 4. Divide by 4 and the norm is an integer. All the norms map back to Z.

An element is a unit iff its norm is a unit, iff its norm is 1. The group of units includes the 8 we saw before, ±1, ±i, ±j, and ±k, plus another 16 units produced by placing either ½ or -½ in each of the four slots. Thus ½-½i+½j+½k is a perfectly good unit, with norm 1.

Plot these 24 units in 4 dimensional space. They all lie on the unit sphere, all 1 distance away from the origin. The original 8 units are ±1 on the four coordinate axes. The other 16 points lie on the main diagonals of the 16 regions walled off by the four hyperplanes.

If x consists of half integers, some associate of x consists of integers. Multiply x by a unit u with half integer coefficients, and if that does not do the trick, that is, if ux still shows half integers, then multiply x by u+1 or u-1, whichever is a unit. The "other" unit then adds or subtracts x to the product, which then creates integers. If the real number coefficient of u was ½, you would multiply by u-1. If the real number coefficient of u was -½, you would multiply by u+1. Note that you are simply switching the sign of the real number coefficient of u.

Projective Space and an Interesting Homomorphism

Euclidean space consists of real coordinates in 1, 2, 3, or more dimensions. For example, 3 space is typically plotted with x y and z coordinates, using the x y and z axes to anchor the system. This is denoted R^3, for real numbers cubed.

Projective space consists of lines that pass through the origin. Each line is a single element or a single "point. It is convenient to intersect these lines with the unit sphere. Each line intersects the sphere in two points that are a diameter apart. Conversely, two antipodal points determine a line that passes through the origin. The z axis intersects the sphere in the north and south pole, and the x axis intersects the sphere in two points on the equator. Projective space is really a sphere with opposite points sewn together. Travel halfway around the world and you are back where you started, because opposite points are the same.

The unit sphere in 3 space is denoted S^2, because it is a 2 dimensional surface. If you are a fly sitting on the sphere, it looks 2 dimensional. Projective space is denoted P^2, because it also looks 2 dimensional. If you are a fly on the surface of projective space, you can't really tell that your little patch of ground is the same as the patch of ground on the other side of the Earth. It just looks like a flat patch of ground.

The projective circle in the complex plane is the set of points with norm 1, that are a distance 1 away from the origin, such that s and -s are the same. The x axis ties 1 and -1 together. The y axis ties i and -i together. The line y = x connects +sqrt($\frac{1}{2}$),+sqrt($\frac{1}{2}$)i with -sqrt($\frac{1}{2}$),-sqrt($\frac{1}{2}$)i.

We can do the same thing in the quaternions. The points with norm 1 define S^3, a 3 dimensional sphere living in 4 space. Equate each point with its opposite to get P^3, 3 dimensional projective space living in 4 space. The 4 axes connect 1 and -1, i and -i, j and -j, and k and -k. Each line through the origin connects opposite points on the sphere, such as $\frac{1}{2}+\frac{1}{2}i+\frac{1}{2}j+\frac{1}{2}k$ and $-\frac{1}{2}-\frac{1}{2}i-\frac{1}{2}j-\frac{1}{2}k$.

Surprisingly, this structure is isomorphic to the orthonormal 3 by 3 matrices. P^3 is the same as the rigid rotations of an object about its center in 3 space. The homomorphism is presented below. The input is a quaternion s+ti+uj+vk with norm 1. The output is a matrix.

$s^2+t^2-u^2-v^2$ -2sv+2tu   2su+2tv

2sv+2tu $s^2-t2+u^2-v^2$ -2st+2uv

-2su+2tv   2st+2uv $s^2-t^2-u2+v^2$

Dot each row with itself and get $(s^2+t^2+u^2+v^2)2$, which is 1. Dot any two different rows and get 0. The matrix is orthonormal.

The determinant is $(s2+t2+u2+v2)^3$, or 1. The matrix is a rotation, not a reflection.

Even more surprising, you can multiply two quaternions and place the result in this matrix, or you can map the two quaternions into two orthonormal matrices and multiply them together; the result is the same. Thus we have a homomorphism from the quaternions with norm 1 into the rigid rotations in 3 space.

Negate a quaternion and apply this map; the result is the same. Every term is a product of two variables, so when you multiply each by -1 there is no change. Opposite points on the sphere lead to the same matrix. The homomorphism is well defined on projective space.

/* norm is a homomorphism on quaternions that respects multiplication */

t = "a×e-b×f-c×g-d×h"^2

t = t + "a×f+b×e+c×h-d×g"^2

t = t + "a×g+c×e+d×f-b×h"^2

t = t + "a×h+d×e+b×g-c×f"^2

u = "(a^2+b^2+c^2+d^2)" × "(e^2+f^2+g^2+h^2)"

Calculate t-u and observe that it is equal to zero.

Using the program shown above, quaternions with norm -1 map to orthonormal matrices with determinant -1. It's the same homomorphism; we have simply changed the domain and range. One projective sphere maps to rotations, and another maps to reflections. However this can't happen in real space, because the norm of a quaternion is always positive; thus the homomorphism is typically restricted to quaternions with norm 1.

This map is injective, an embedding, a monomorphism (all words for the same thing). Each point in the domain maps to a unique matrix in the range. It is enough to show the kernel is 1. Assume the image is the identity matrix, and set 2st+2uv = -2st+2uv = 0. Thus s or t is 0, and u or v is 0. In the same way, s or u is 0, and t or v is 0 - and s or v is 0, and t or u is 0. If two variables are nonzero, one of these 6 constraints fails. Thus three of the four variables are 0. We are talking about 1, i, j, or k. Only 1 produces the identity matrix, hence the map is an embedding. (-1 also produces the identity matrix, but -1 and 1 are the same thing in projective space.)

Let's complete the isomorphism for the reals. We only need show the map is onto.

The bottom row of a rigid rotation is the image of the z axis, the destination of the north pole. The top row is the destination of the x axis, which defines a rotation about the new location. Picture this as an arrow tangent to the sphere. We need to move the north pole to every location, and then point the arrow in every direction.

Set t = u = 0 and let s and v spin around the unit circle. The bottom row is [0,0,1], and the right column is [0,0,1]. This fixes the north pole. Think of s as cos(theta) and v as sin(theta). The corresponding rotation maps x to cos(-2theta),sin(-2theta), via the double angle formulas, and y runs 90 degrees ahead. The quaternions make the arrow spin all the way around the north pole. In fact the arrow spins twice as fast, but we expect this when the domain is projective space, since opposite points on the sphere are the same. Run half way around the sphere and you are back to start; and the rotation, represented by an arrow at the north pole, is back to start.

Now for something fairly intuitive. Let the arrow spin around the north pole, then move the north pole to a location l on the sphere. Compose the two functions and the arrow spins around l. Choose your frame of reference so that 0 degrees at the north pole corresponds to 0 degrees at l. Moving to l is a linear map, and it even preserves angles. An arrow at 17 degrees becomes an arrow at 17 degrees. All directions are covered; we only need map the north pole to l.

Set t and v to 0, and let s and u spin around the unit circle. This carries the north pole to any point on the meridian around the y axis, in the xz plane. Use this to move the north pole to any latitude; then apply a rotation about the north pole to move this point to any longitude. That covers the entire sphere, and the map is onto.

The matrix functions are quadratic in s t u and v, hence continuous. The spaces are compact and hausdorff, hence the map is a homeomorphism. Compact means closed and bounded. Hausdorff means that individual points are separated. "Homeomorphism" is topological terminology to say the spaces are the same. You can't really tell if you live in one space or the other, so use whichever space you like.

It is difficult to analyze the rotations in 3 space, i.e. the arrows on the sphere, but projective space is well understood. The homotopy and homology groups of projective space carry over to rotational space for free. Of course this is a book about discrete structures, not continuous maps between topological spaces; but I just had to include this result. It's just too beautiful!

When applied to other fields, this map is not always surjective. Rational quaternions lead to rational rotations, but not all rational rotations are covered. Spin the arrow around the north pole 90 degrees, which has the following rational matrix.

0	1	0
-1	0	0
0	0	1

Embed everything in the reals and this rotation can have only one preimage, which is based on the sine and cosine of 45 degrees. These do not have rational coordinates. The map is still a monomorphism, but no longer an isomorphism.

The Square Root of a Rotation

Every rotation in 3-space has a square root. If S is the orthonormal matrix that implements the rotation, there is some other rotation R such that R×R = S. Apply R twice to get S.

This is clear in 2 dimensions. If S spins the plane through an angle of theta, then R spins the plane through an angle of theta/2, or theta/2+p.

I guess it's pretty clear in 3-space as well, as long as you know that every rotation has an Eigen vector, an axis of rotation, a line that stands still, like the north pole. Show that R and S have to have the same axis of rotation. From there, R is a rotation about this axis, through an angle of theta/2, or theta/2+p.

This gives us a warm fuzzy feeling; every rotation has exactly 2 square roots. However, there is one exception. The identity rotation, where the Earth stands still, has infinitely many square roots. Spin the Earth about any axis 180 degrees, and that is a square root of 1.

What if the base field is different than the reals? What if we are working with the integers mod p, and there is no axis of rotation? Pull the rotation back to a quaternion using the isomorhism described in the previous section, take its square root, and push it forward to find the desired rotation. The square root of the quaternion may require a field extension, like adjoining i to the reals, but magically this goes away when you turn it back into a rotation.

Assume the rotation corresponds to the quaternion $s+ti+uj+vk$, and set this equal to $(w+xi+yj+zk)2$. Equate the imaginary terms, and $2wx = t$, or equivalently, $x = t/2w$. Similarly, $y = u/2w$, and $z = v/2w$. Now the real term becomes $w2 - (t2+u2+v2)/4w2$. The norm is 1, so rewrite this as $w2 + (s2-1)/4w2$. This must equal s.

$4w4 - 4sw2 + s2-1 = 0$

The discriminant is 16, and $w^2 = (s\pm1)/2$. Extend the field F if necessary, so that w is well defined. Set aside for the moment the case of $s = 1$ or $s = -1$, corresponding to the identity matrix, so there is no danger of $w = 0$. We are taking the square root of a nontrivial rotation.

For $s+1$, the norm of $w+xi+yj+zk$ is 1. For $s-1$, the norm is -1, and that's not what we want. The other rotation comes from the square root of $-(s+ti+uj+vk) = w - (ti+uj+vk)/2w$, where $w^2 = (1-s)/2$. So $w^2 = (1+s)/2$ or $(1-s)/2$.

It looks like there are 4 square roots, but negating w negates x y and z as well, and that is the same point in projective space, so there are actually 2 square roots as expected, corresponding to 1+s and 1-s.

Map $w+xi+yj+zk$ forward and get this matrix. There is a common denominator of $4w^2$, or $2(1\pm s)$, which I did not include in each entry. With this in mind, the effect on the matrix is to replace every s with $1\pm s$, negating t u and v for the 1-s case.

Here is the 1+s case.

$(1+s)^2+t^2-u^2-v^2$	$-2(1+s)v+2tu$	$2(1+s)u+2tv$
$2(1+s)v+2tu$	$(1+s)^2-t^2+u^2-v^2$	$-2(1+s)t+2uv$
$-2(1+s)u+2tv$	$2(1+s)t+2uv$	$(1+s)^2-t^2-u^2+v^2$

Here is the 1-s case.

$(1-s)^2+t^2-u^2-v^2$	$2(1-s)v+2tu$	$-2(1-s)u+2tv$
$-2(1-s)v+2tu$	$(1-s)^2-t^2+u^2-v^2$	$2(1-s)t+2uv$
$2(1-s)u+2tv$	$-2(1-s)t+2uv$	$(1-s)^2-t^2-u^2+v^2$

Although W may not lie in our base field F, the above matrix does. Verify, by computer, that this matrix is orthonormal, with determinant 1, and its square is the original matrix in the s t u v format we saw before. This is the 1+s case.

Higher dimensions have more square roots. Let a 4 by 4 matrix be block diagonal, with a 2 by 2 block in the upper left and a 2 by 2 block in the lower right. These are two 2-dimensional rotations in parallel, and there are at least four square roots.

The Hairy Ball Theorem

Speaking of arrows on the sphere, here is a theorem that belongs squarely in the continuous camp, so you can skip it if you like.

It is not possible to comb the hair on a ball continuously, without creating a whorl, where the hair spreads out in different directions, like the top of your head, or a cowlick, or a point that has no direction because the hair swirls around it like the eye of a hurricane. In other words, the tangent vector has to be 0 somewhere on the ball. You just can't smoothly comb the ball all the way around.

You can of course comb the hair around a circle, in one of two directions, clockwise or counterclockwise. You only get into trouble on the sphere. A more general proof addresses spheres of all dimensions, but this proof is specific to S^2 .

Put the sphere in front of you with the north pole at the top and the south pole at the bottom. Let arrows represent the tangent vectors, i.e. the direction of the hair. Walk around the equator heading east, always looking inward toward the center of the sphere, and note the direction of the arrows as you go. The direction moves continuously. It may jiggle and jitter about, and even spin around, but when you get back to start on the equator, the arrow has returned to its original direction. This is a continuous map from the circle (equator) into the circle (arrow direction).

Let the winding number be the net number of times the arrow spins around clockwise as you travel around the equator. This has to be an integer. If the arrows point east, for example, all the way around the equator, the winding number is 0. If the arrows point north all the way around, the winding number is still 0. More likely, the arrows will move this way and that as you go, and perhaps create a net spin when you get back to start.

This can be done for any latitude, other than 90 or -90 degrees, which are degenerate cases. If l is the latitude, walk around the sphere at latitude l and note the winding number of the arrows at that latitude. Call this w(l).

The arrows are continuous across the surface of the sphere, and that means they change hardly at all between 51.23 degrees and 51.24 degrees, for example. More formally, w(l) is continuous in l. Since w(l) is an integer, it can't continuously jump from one number to another. Therefore w(l) is constant across all latitudes.

Continuous on a compact set is uniformly continuous, so there is some little disk about the north pole, say 89 degrees north and above, where the arrows don't vary by more than 1 degree. They are all pretty much in line. If the arrow at the north pole points right then everything above 89 degrees points right. This determines w(l). March around the circle at 89 degrees north. First the arrows point right, which is east, which is the direction of travel. But a quarter of the way around the circle the arrows point south, relative to the circle being traced.

On the back side the arrows point west, then they point north, then east again. The arrows spin around once clockwise, hence w(89) = 1, hence w(l) = 1.

Do the very same thing at the south pole, and w(l) = -1. The arrow spins around once counterclockwise as you traverse the equator. Since w(l) cannot be both 1 and -1 simultaneously, the continuous field of arrows cannot exist.

If you delete the south pole then the problem goes away. w(l) can equal 1 for -90 < l < 90.

With w(l) = 1, as when covering the northern hemisphere, the arrow must spin once around (net) as you travel the equator. There is some point on the equator where the arrow points directly south, and another where the arrow points directly north. Flatten this out, and a continuous field of arrows on the closed disk can exist only if some arrow, somewhere on the circumference, points straight in, and another arrow points straight out.

Generalized Euclidean Space

Take the direct sum of arbitrarily many copies of the real line, using zero as the preferred element. Every point in this composite space has finitely many nonzero coordinates; all the other coordinates are 0.

Call this space E^j, for Generalized Euclidean Space. By default, E^j joins a countable collection of real lines together, but higher cardinalities are possible. Of course, if you join a finite number of real lines together you simply get R^n, i.e. n dimensional space.

-3	-2	-1	0	1	2	3

×

-3	-2	-1	0	1	2	3

×

-3	-2	-1	0	1	2	3

×

…

This is a real vector space. Add two points together by adding their coordinates. Scaling by a real number c multiplies all the coordinates by c.

Construct a distance metric on Ej as follows. Take any two points and note that their coordinates agree almost everywhere. In fact their coordinates are 0 almost everywhere. Subtract the coordinates that disagree, square the differences, add them up, and take the square root. Since two points can only disagree on a finite number of coordinates, this is well defined. In fact it is just the Pythagorean theorem in n-space.

Take any three points and derive their distances, using the above formula. Note that all three points live in Rn for some integer n. Furthermore, the distance metric is identical to distance in n-space. Therefore, the triangular inequality holds. You can't build a triangle where one side is longer than the sum of the other two sides. The metric is valid across all of E^j. This structure is a vector space, and a metric space, with a topology of open and closed sets. Such a structure is called a topological vector space, and I may explore these in a later chapter.

Polynomials

A polynomial is basically a sequence a_n … a_3 a_2 a_1 a_0, that is, by convention, written as $a_n x^n + … + a_3 x^3 + a_2 x^2 + a_1 x + a_0$. The polynomial takes on a special significance when you substitute something in for x, but sometimes the polynomial has a life of its own.

To be useful, polynomial arithmetic should commute with substitution. What does this mean? If p(x) and q(x) are two polynomials, we can add p and q, then substitute for x, or we can replace x in both polynomials and add the two values. The answer should be the same. You can't add $x+x^2$ and get x^3, for then 2+4 would equal 8. After some thought, and perhaps a really tedious proof, you will convince yourself that polynomials must be added term by term, and multiplied by distributing terms across terms. I will assume you know the procedures for handling polynomials.

A polynomial typically has no inverse. A constant polynomial might be invertible, like 7 in the reals, with inverse 1/7, or the polynomial 2x+1 mod 4, which is its own inverse, but these are special cases. If you want inverses you usually resort to fractions, like $(x^2+2x+7) / (3x+5)$. These are called rational functions, just like fractions of integers are called rational numbers. The rational function 1 over (x-1) is a perfectly good function, but it becomes undefined if you replace x with 1, because the denominator drops to 0.

Polynomials can contain more than one variable, such as $2x^2 + 7xy + 5y^2 + 3x + 1$. It is sometimes helpful to view this polynomial as a polynomial in x, whose coefficients are polynomials in y, or a polynomial in y, whose coefficients are polynomials in x. This continues the theme of objects within objects.

$2x^2 + (7y+3)x + (5y^2+1)$

$5y^2 + 7xy + (2x^2+3x+1)$

Unless stated otherwise, x and y commute. Thus $(x+y)^2 = x^2+2xy+y^2$. However, if they do not commute, then $(x+y)^2 = x^2+xy+yx+y^2$, and xyxyx cannot be simplified to x^3y^2. Furthermore, we may not be able to rewrite the expression as a polynomial in x, whose coefficients are polynomials in y, as illustrated above. That might be off the table.

Synthetic Division, Roots, and GCD

Synthetic division divides one polynomial into another, giving a quotient and a remainder. The process is very much like long division. It works as long as the lead coefficient of the divisor is invertible. Let's divide 2x+4 into x3+5x+1. Look at the lead terms and divide x^3 by 2x. (I'll assume we are working in the rationals; this can't be done in the integers.) Thus, the quotient begins with $(\frac{1}{2})x^2$. This is like the first digit of the quotient in long division. Multiply this by the divisor and subtract from the dividend, giving -2x2+5x+1.

Take the next step and divide 2x into $-2x^2$, giving -x. Multiply and subtract, leaving 9x+1.

The last "digit" is 9/2, and the quotient is $\frac{1}{2}x^2$-x+9/2, with a remainder of -17.

Assume (x-r) is a factor of h(x). In other words, (x-r) ×g(x) = h(x). Substitute r for x and h(r) = 0. That means r is a root, or zero, or solution, of h(x). Conversely, let h(r) = 0 and divide x-r into h(x) by synthetic division. We can always do this because the lead coefficient of (x-r) is 1. The remainder has to be a constant c. Thus h(x) = (x-r) ×g(x) + c. Substitute r for x, and c has to be 0. Therefore, r is a root of h(x) iff x-r is a factor of h(x).

Note that (x-r) could appear more than once. For instance, x^2-2x+1 has only one solution, namely 1, but this root has multiplicity 2, because $(x-1)^2 = x^2$-2x+1.

A wonderful corollary is that a polynomial of degree n cannot have more than n different roots. Suppose p(x) has degree 6, that is, it starts with x^6, yet it has roots 1, 3, 5, 7, 9, 11, and 13. Divide p by x-1 and call the quotient q. Thus p = (x-1) ×q(x), and q has degree 5. Replace x with 3 and q(3) has to equal 0. Thus 3 is still a root of q. Replace q with (x-3) ×r(x). Now 5 is a root of r. Continue this process until p = (x-1) × (x-3) × (x-5) × (x-7) × (x-9) × (x-11). Great, but there's still another root. Plug in 13, and the product is nonzero; yet 13 is suppose to be a root of $p(x)$. This is a contradiction, hence there are at most n roots for an n degree polynomial.

This works as long as multiplication is commutative. The equation x^2+1 has at least 6 roots in the quaternions, namely $\pm i$, $\pm j$, and $\pm k$. What goes wrong when multiplication is not commutative? Since i is a root, write $p(x) = (x-i) \times (x+i)$. Plug in the next root j and get $(j-i) \times (j+i)$. This becomes $j^2-ij+ji-i^2$, or $-2k$. The polynomial arithmetic does not match the quaternion arithmetic. We started with $(x-i) \times (x+i) = x^2+1$, assuming x and i commute, but after substitution they do not commute. Synthetic division (pulling out x-i) does not give a proper quotient.

Even if multiplication is commutative, and synthetic division is valid, there is one more thing to watch out for. After finding 6 of the 7 roots, I multiplied 6 nonzero factors together and assumed the result was nonzero. In other words, the theorem requires no zero divisors. To see what happens when zero divisors are present, consider the integers mod 8. Mod 8 has plenty of zero divisors. The polynomial x2-1 has 4 roots mod 8. With 1 and -1 as roots, write x^2-1 as x-1 times x+1, then replace x with 3 or 5. The product still comes out 0.

Synthetic division also supports Euclid's greatest common divisor (gcd) algorithm, provided all coefficients are invertible. Consider polynomials over the rationals, so that every coefficient is invertible.

Let $f(x)$ be a factor of both $a(x)$ and $b(x)$. In other words, f goes into a and b evenly. I will often write this as f divides a, and f divides b. When "divides" is a binary operator, just think of it as "goes into", as you learned in elementary school. It is an assertion of divisibility - unlike the operator / or $\div$, which generates a quotient.

Divide b into a. This gives quotient q and remainder r. Thus $a(x) = b(x) \times q(x) + r(x)$. Now f is a factor of r. In other words, f divides r. Conversely, if f divides r and b then f divides a. The same set of polynomials divides b and r, as a and b, so set a = b and b = r and start again. The degree decreases with each step, until the remainder is 0. At this point the divisor, call it $g(x)$, is the greatest common divisor, because g divides a and b, and every f that divides a and b also divides g. g is the highest polynomial that is common to a and b.

Try it with x^3-1 and x^2+x-2. These are a and b; use synthetic division to divide b into a, giving a remainder of 3x-3. Since 3 is a unit in Q, go ahead and pull 3 out of 3x-3. This won't change the polynomials that divide another polynomial. Now $a = x^2+x-2$ and $b = x-1$. Since b divides a with no remainder, x-1 is the gcd. This is confirmed by 1, which is a root of both x^3-1 and x^2+x-2.

The gcd algorithm is efficient, and runs on polynomials with thousands of terms. If the coefficients don't explode into ungainly fractions, the algorithm runs in n^2 time.

There is a sense in which a polynomial can be prime or composite. Consider $p = x^2+1$, over the integers, or rationals, or reals. If it factors it has to split into two linear pieces. Their product is x^2+1. Their lead coefficients multiply to 1. Over the integers these lead coefficients are 1 or -1; if -1 then negate each factor. Over the rationals or reals, one coefficient is the inverse of the other. Perhaps one is 7 and the other is 1/7. Again, multiply through so that the lead coefficients are both 1. Now each factor looks like x-r, and r is a root. But x^2+1 has no solution. Therefore x^2+1 does not factor, and is prime.

Consider x^3+x+3 over the integers. If it factors then one of the two factors is linear and the other is quadratic. (The quadratic could split again but we'll deal with that later.) If the lead coefficients are -1 then negate the two factors, so that the lead coefficients are 1. The linear factor is x-r, and r is a root. Multiply the two factors together and r is an integer that divides 3. Try ±1 and ±3; but none of these are solutions, so x^3+x+3 is prime.

However, $x^4 + 4$ is not prime. It factors as $(x^2 - 2x + 2) (x^2 + 2x + 2)$

Most polynomial rings exhibit unique factorization; each polynomial splits uniquely into primes. This will be proved later.

Even a multivariable polynomial could be prime, if it does not factor. Recall our earlier example, which we wrote as a polynomial in x, whose coefficients are polynomials in y.

$2x^2 + (7y+3)x + 5y^2+1$

Try to factor this over the integers. This is quadratic relative to x, and the three coefficients have nothing in common, so if it factors the two factors are linear in x. Multiply by -1 if necessary, so that both lead coefficients are positive. One is 1 and the other is 2. The product of the constant terms is 5y2+1, which is prime (as a polynomial in y), so our factor looks like x, or 2x, ±(5y2+1), or ±1. There are eight possible roots: ±(5y2+1), or this divided by 2, or ±1, or ±½. The first four create a polynomial in y of degree 4, which is clearly not 0. The remaining four leave 5y2 hanging around. None of these work, hence the polynomial in x and y is prime.

Formal Derivative

Inspired by calculus, the derivative of a polynomial is produced by folding the exponents into the coefficients, and decreasing the exponents by 1. Thus $4x^3 +5x+1$ yields $12x^2 +5$. This is called a formal derivative because it is not used to compute the rate of change. It is strictly algebra. If the coefficients are taken mod m, then the derivative of 7xm is 0, simply because 7m = 0.

Verify that the derivative of p(x)+q(x) is the sum p' + q', and the derivative of cp(x) is c times p'. This follows directly from the definition.

Again, from calculus, let the derivative of pq be p'q + pq'. Is this consistent with everything that has gone before? Apply this to x^k times x and get $k \times x^{(k-1)} x + x^k \times 1 = k \times x^k +x^k = (k+1) x^k$. Also apply this to c times p(x) and get $c \times p'(x)$. This because the derivative of a constant c is 0. So far so good.

Apply the product rule to p × q where p and q have many terms. A typical term in the product, multiplying cx^k by dx, is cdx^kx, with a derivative of $cd(k+l)x^k$. Now apply the product rule and get $cx^kdx^0 + ckx^{k-l}dx$. It's the same.

Using induction on n, and the product rule, prove that the derivative of $p(x)^n$ is n times $p(x)^{(n-1)}$ times $p'(x)$.

$$(pn)' = p \times (p^{(n-1)})' + p' \times p^{(n-1)}$$

Spread this across an entire polynomial and get the chain rule. The derivative of $q(p)$ is $q'(p) \times p'$.

Formal derivatives can extract repeated roots or repeated polynomial factors. For instance, $x^3 + x^2 - 5x + 3 = (x-1)^2 \times (x+3)$. We can extract the repeated factor $(x-1)$, which implies a repeated root of 1.

Here is how you do it. Consider the derivative of qs^n, where q and s are polynomials and $n > 1$. Apply the product rule and the chain rule, and the derivative is divisible by $s(n-1)$. Take the gcd of the original polynomial with its derivative to extract $s(n-1)$. The factors with high multiplicity fall out.

Conversely, assume the polynomials exhibit unique factorization, and s is a prime polynomial. Let both h and h' contain $s^{(n-1)}$. We know that h does not include $s(n+1)$, else h and h' would both contain s^n. Suppose h contains fewer than n instances of s. I'll illustrate with n-1. Write $h = qs^{(n-1)}$ and apply the product rule for differentiation. We must account for n-1 factors of s, hence s divides qs'. Since s' has a smaller degree than s, s divides q. This contradicts our assumption. Therefore h has exactly n instances of s.

There is a catch; if s' drops to 0, then s does not have to divide into q after all. This is rarely a problem, even in modular mathematics. I'll address this possibility if and when it emerges. In general, h contains s^n iff h and h' both contain $s^{(n-1)}$.

Power and Laurent Series

A power series is basically an infinite sequence $a_0\ a_1\ a_2\ a_3 \ldots$ that happens to be written $a_0 + a_1x + a_2x^2 + a_3x^3 + \ldots$. A polynomial starts with x^n at the left, but since a power series goes on forever, we turn it around and start with the constant term and let the exponents increase from there. You will eventually encounter some dots, or a formula for a^n, because you don't have an infinite amount of ink.

Substitution is a little trickier, since the series might not converge. Let each $a_i = 1$, and the series is $1/(1-x)$. (Divide $(1-x)$ into 1 by synthetic division and see.) This converges only for x between -1 and 1.

When $a_n = 1/n!$ the series is E^x, which converges everywhere. The power series is the Taylor series for E^x.

The series becomes a polynomial if everything after a_n is 0. That means series arithmetic is backward compatible with polynomial arithmetic. Series are added term by term, and multiplied in a diagonal fashion. Like the product of two polynomials, $\{c_n\} = \{a_n\} \times \{b_n\}$, c_n is the sum of $a_ib_{(n-i)}$, as i runs from 0 to n.

Synthetic division works, but is often impractical, since each intermediate result is infinitely long. Sometimes there are shortcuts. You could compute the terms of 1 over E^x by dividing the power series for Ex into 1, or you could note that 1 over Ex is E(-x), whence the series is $(-1)^n/n!$. Multiply this by the series for Ex and look at c_n. Let $j = n-i$, and write c_n as the sum of $1/i!$ times $(-1)^j/j!$. Multiply top and bottom by n! and pull out the common denominator of n!. You are left with $(1-1)^n$, as expanded by the binomial theorem. This is 0. Each c_n is 0, except for n = 0, so yes indeed, E^x times $E^{(-x)} = 1$.

A Laurent series can start with any negative exponent. Let $a_i = 1$ for i = -2. This is the series for 1 over $(1-x)x^2$. This converges for any x between -1 and 1, other than 0.

Series are added term by term, and multiplied in a diagonal fashion, starting with the first term in the series, which could have a negative exponent. If the least exponent in two series is s, then cn, in the product, is the sum of $a_i b_{(n-i)}$, as i runs from s to n-s.

As you have probably guessed, the power series and the laurent series are inspired from their counterparts in calculus. These describe analytic and meromorphic functions respectively. Yet a power series can be a useful algebraic tool outside the context of calculus, just as the formal derivative can be used to search for repeated roots of a polynomial.

A series can be multivariable, including both x and y for example. By convention we start with the constant, then the terms of degree 1, then degree 2, then degree 3, etc.

$3 + 2x - 5y + 7x^2 + 11xy - 5y^2 - 36x^3 \ldots$

Sometimes such a series can be rewritten as a series in x, whose coefficients are series in y, but sometimes the terms cannot be drastically rearranged in this manner.

p-adic Numbers

The p-adic numbers are a bit like a power series with coefficients mod p, but there are carry operations. I'll illustrate with p = 7.

Write all the integers in base 7, but reverse the digits, so that 120 comes out 132 rather than 231. The first digit is multiplied by 1, the next by 7, the next by 49, the next by 343, and so on. You could write this as $1 + 3x + 2x^2$, or $1 + 3p + 2p^2$, with the understanding that x, or p, is replaced with 7, but it's easier to just write 132, which begins to look like a number in base 7. Just remember you are looking in a mirror.

Two strings are added together digit by digit, with carry operations that flow to the right, rather than to the left. Thus $5p + 4p = 2p + p^2$, or 05 + 04 = 021. This is different from polynomials mod 7, where 5x + 4x = 2x.

Multiplication also takes place in the usual manner, the way you learned in grade school, but it is in base 7, and reflected. Here is an example: 213 × 411. Does this remind you of Tom Lehrer doing New Math? "Base 8 is just like base 10 really, if you're missing two fingers."

411

1551

213

213

=10263

Here's the twist; infinite strings of digits are allowed. These do not correspond to integers, with a few exceptions. Consider the string 6666666… forever. Add 1 and let the carry operation ripple to the right. Everything turns to 0, hence this string is -1. Similarly, 356666666… is -11. Ad 41000000… and get 0. Other than these negative numbers, infinite strings are something else entirely.

Some of them are fractions. Multiply 3 by 5444444… and get 1, hence 5444444… is 1/3. Repeating sequences correspond to fractions, just as repeating decimals correspond to fractions in the reals. This is a consequence of synthetic division. However there are many more sequences, analogous to irrational numbers.

Series are added digit by digit, carrying to the right, and multiplied in a diagonal fashion, with overflow pushing to the right. The set of all these strings, under base p arithmetic, is called the p-adic numbers.

There is a homomorphism from the p-adic numbers onto the integers mod p. Just look at the first digit and ignore the rest. The first step in adding or multiplying two series is to add or multiply their first digits, hence the map respects addition and multiplication.

In general, a homomorphism takes the p-adic numbers onto the integers mod p^k, by considering only the first k digits. Everything else is thrown away.

This is not a complete field of fractions. The integer 7, which is represented 01, has no inverse. You can't multiply this by anything to get 1 in the first position. However, a laurent series fills in the gaps. This brings in negative exponents, such as 1/p. In fact, 1/p times p = 1. If a vertical bar indicates the start of the power series, a bit like a decimal point, you could write this as: 1|00 × |01 = |1. The inverse of 49, or |001, is 10|0, and so on. You can always divide one laurent series into another, so everything is invertible. This embeds the p-adic numbers into their fractions.